SWITCHING GEARS

Published by River Grove Books
Austin, Texas
www.rivergrovebooks.com

Distributed by River Grove Books

For ordering information or special discounts for bulk purchases, please contact Greenleaf Book Group at PO Box 91869, Austin, TX 78709, 512.891.6100.

Design and composition by Greenleaf Book Group
Cover design by Greenleaf Book Group

Cover image ©Angelatriks. Used under license from Shutterstock.com.

For permission to reproduce copyrighted material, grateful acknowledgments is made to the following sources:

From "That Tesla Battery Emissions Study Making the Rounds? It's Bunk" by Ezra Dyer, from *Popular Mechanics,* June 22, 2017. Copyright © 2017 by Hearst Publications. Reproduced by permission of Hearst Publications. http://www.popularmechanics.com/cars/hybrid-electric/news/a27039/tesla-battery-emissions-study-fake-news/

From "Bad Karma: Fisker's Government-Funded Failure" from Consumer Reports, March 8, 2012. Copyright © 2012 by Consumer Reports. Reproduced by permission of the copyright holder. https://www.consumerreports.org/cro/news/2012/03/bad-karma-our-fisker-karma-plug-in-hybrid-breaks-down/index.htm

Map "Registration and Title Fees by State." Copyright © 2018 by National Conference of State Legislatures. Reproduced by permission of the copyright holder.

http://www.ncsl.org/research/transportation/registration-and-title-fees-by-state.aspx

Publisher's Cataloging-in-Publication data is available.

Print ISBN: 978-1-63299-323-6

eBook ISBN: 978-1-63299-324-3

First Edition

DAN K. EBERHART

SWITCHING GEARS

THE PETROLEUM-POWERED ELECTRIC CAR

RIVER GROVE
BOOKS

CONTENTS

INTRODUCTION

On a Midwestern morning in 1890, a young chemist by the name of William Morrison rolled the first successful "electric car" out of his shop in Des Moines, Iowa. Little more than a six-passenger electrified wagon, this vehicle could cruise at a top speed of 14 miles per hour.

At the time, motorists in the United States had essentially three options if they wanted to upgrade from their old nag: steam, gasoline, or electricity. Steam, the energy source that had been powering factories and propelling trains for years, was foraying into self-propelled vehicles. Gasoline had gained popularity as the technology of internal combustion engines improved. And electricity was becoming easier for average consumers to access.

Electricity soon became the energy source of choice for many new drivers, who either didn't feel like waiting the long time required to start up a steam engine or had a hard time hand-cranking the gasoline engine. (Here's where the roadster's nickname "runabout" came from: The driver had to crank the handle to start the car, then *run about* the car while avoiding the puff of smoke emanating from the engine.)

The four-horsepower 1890 Morrison Electric was powered by twenty-four storage battery cells, which Morrison had designed himself. The batteries, mounted under the front seat, could produce 112 amperes at 58 volts. The whole setup took ten hours to recharge.[1] And it's reported that the Morrison could run up to one hundred miles without recharging.

Soon, demand for electric vehicles (EVs) grew—especially among city dwellers, whose access to electricity was growing. Just before the turn of the century, New York City pedestrians could hail more than 60 electric taxis. And by 1900, electric cars made up a third of all vehicles on the road across the nation.

Around the same time, the new Ford Motor Company introduced the assembly line, an efficient innovation that streamlined automobile production and lowered costs. By 1912, average car costs had really begun to diverge: The average internal combustion engine (ICE) car cost $650, while the average electric roadster price tag was more than twice as much at $1,750.[2]

That wasn't the only thing to diverge. While roadways continued to improve, connecting urban centers and winding through sparsely populated small-town U.S.A, gasoline filling stations sprang up like weeds along newly paved highway systems. Meanwhile, access to electricity outside major urban centers remained limited, with no significant growth expected.

And at this metaphorical fork in the road, ICEs sped into the lead, leaving their electric counterparts in the dust and without a charge in sight. Production of EVs stalled out as consumers opted in increasing numbers for the more affordable and reliable ICEs. By 1935, William Morrison and the earliest EVs had faded into obscurity.

The EV story was over. Or so we thought.

In 1997, as worldwide concerns about emissions grew, Japan rolled out the Toyota Prius hybrid. Two years later, in 1999, the Prius came to the United States. Other vehicle manufacturers at home and abroad followed suit. And then the 2006 San Francisco International Auto Show saw a young Silicon Valley–based company called Tesla Motors unveil the Roadster, the first commercially produced automobile powered by a lithium-ion battery.[3]

Of course, one vehicle does not a revolution make—especially given the high cost of entry: The first Roadster was priced at nearly $100,000—well beyond reach of the average American. By 2009, though, most major car manufacturers had leapt into the fray, spending billions on designing more energy- and battery-efficient EVs in anticipation of big market demand. Companies like Nissan, Chevy, and Honda released EVs aimed at middle-class buyers.

Soon after that, the first EV "battery wars" began, pitting the hydrogen fuel cell against the superconducting lithium battery, and the aluminum-air battery against the new graphene battery, all vying to be the prominent technology for new vehicles. The still-experimental graphene technology uses coatings of very fine carbon on battery poles to boost electrical conductivity, which could potentially increase power production for electric vehicles, as will heavy-duty aluminum-air batteries that utilize aluminum-coated anodes.

Today, EVs are slowly gaining acceptance. They evoke promises of new energy sources and cleaner vehicles. They are the holy grail of energy solutions: power without fossil fuel—or so it appears.

The transportation industry's new rock stars are radical entrepreneurs with visions that could change the landscape of energy as radically as computers have changed the landscape of communication.

The future has never seemed more like science fiction. We've seen hydrogen fuel-cell-powered trains ("hydrail"), autonomous drones, and the first prototypes and working models of electric jets and vertical takeoff and landing (VTOL) vehicles. The people's taxi company, Uber, vows that its Elevate program will lift intercity EVs to the sky,[4] and smaller startups have demonstrated ingenious contraptions for human-powered flight.

The world is on a precipice of energy innovation. Which way will the cards fall? As we strive toward cleaner fuels, which technologies will rise to prominence, leaving competitors as mere sidenote filler for Wikipedia stubs and trivia questions? Will the Tesla Roadster and the Nissan Leaf go the way of the Morrison Electric? Will EVs find themselves as the losing tech once again, akin to Beta to VHS or Sega Dreamcast to Sony PlayStation?

Or are EVs back for real, this time?

Certainly, Bloomberg New Energy Finance believes in the EV revolution. It predicts that a whopping 530 million EVs will be hitting the highways by 2040, up from about 2 million EVs today. Morgan Stanley's analysts say EVs will account for 50 to 60 percent of global light-vehicle sales by 2040. Those EVs could displace millions of barrels of fossil fuels.

Yet, even as we continue our journey to our new and improved energy mix, the path forward is complex.

Global GDP will nearly double between 2018 and 2040, as consumption and population grow.[5] The earth is expected to add the equivalent of another China—1.3 billion people—by 2050, making an all-EV world harder to attain. Population growth produces unprecedented levels of trade, commerce, and traffic, adding hugely to the numbers and movement of people and goods around the planet.

Worldwide, petroleum demand is expected to grow by an average of 1.2 million barrels per day each year until 2022. Furthermore, the United States is projected to dominate oil and gas markets for years to come as the shale boom virtually guarantees the biggest supply surge in history.[6]

As the demand for energy grows, many people are turning to EVs to provide cleaner energy options for transportation. However, EV battery technology requires invasive mining of "rare earths." These are seventeen elements so classified in the periodic table:

Cerium (Ce)	Praseodymium (Pr)
Dysprosium (Dy)	Promethium (Pm)
Erbium (Er)	Samarium (Sm)
Europium (Eu)	Scandium (Sc)
Gadolinium (Gd)	Terbium (Tb)
Holmium (Ho)	Thulium (Tm)
Lanthanum (La)	Ytterbium (Yb)
Lutetium (Lu)	Yttrium (Y)
Neodymium (Nd)	

The rare earths are much in demand for magnetic and electronic components in cars, laptops, cell phones, power tools, and all manner of precision digital equipment. The rare earth mining process, as has been amply demonstrated in China, can damage local environments with toxic chemicals (specifically ammonium sulfates and acid baths pumped into wells) as much as or more than digging for coal or fracking oil or natural gas.[7]

In addition, an "emissions-free" EV can never truly be emissions-free, partially because of emissions created through the manufacturing process and certainly if the electricity used to charge it is produced by coal or other conventional sources. What's more, traditional gasoline-powered cars and trucks are becoming much more energy-efficient. Even today, smaller smart cars and clean diesel vehicles already offer 40 miles per gallon (mpg) and higher, as do 4 million hybrid electric-gas vehicles operating in the United States.

Politically, as well, the road to EV adoption is rocky. While those who believe EVs are a clear answer to climate change push for government subsidies, others argue that the government should leave consumers in the driver's seat and let the free market decide. Those against government spending on EVs also point out that although most experts agree lifetime EV emissions are currently lower than their ICE counterparts, the jury is still out as to whether EVs are a *cost-efficient* means for emissions reduction.[8]

Sometimes decisions seem to be more about image than science. No one wants to be seen as a climate enemy, and so countries are swept forward by political tides, often without pausing to carefully consider whether using taxpayer money to fund EVs is a reasonable way to meet emissions goals. Consider that one recent study actually found EV subsidies to cause *increases* in carbon emissions.[9]

Unfortunately, as countries swarm to set dates for banning ICE sales (Norway in 2025; Scotland by 2032; France and the United Kingdom by 2040, to name a few[10]), they seem to be doing so with little regard as to whether these dates are realistic or, in fact, even necessary. And it's interesting to note that, in the face of all these ambitious proposals, none of this passionate drive (pun intended) has yet amounted to official legislation. "There is literally not a single ban on the books in regulatory language that is enforceable in any auto market in the world," Nic Lutsey, director of the International Council on Clean Transportation, pointed out in August 2018.[11]

Free market advocates suggest we let the market choose the winning technologies. Successful low-emissions technologies will be those that are innovated in the true spirit of American capitalism: We should not let the government choose for us which technologies we need, but rather let the entrepreneurs pitch us the best solutions. The market will decide, and it will be good. Just as cell phones did not need government subsidies despite huge infrastructure and habit-changing barriers, the winning automobile solutions will not need subsidies to become popular.

On the other hand, EV subsidy supporters argue that climate change is coming too fast, and because the EV tech is not yet cost competitive with ICEs (EVs are still much more expensive), only government intervention can move us quickly toward an adoption that is necessary to save the planet from ecological catastrophe.

For a successful energy revolution, we must learn from our mistakes, solve our puzzles, and carefully consider our results, as we work toward a future that allows us to be conscientious, powerful, and energy-savvy all at the same time.

Jules Verne said, "Science, my lad, has been built upon many errors; but they are errors which it was good to fall into, for they led to the truth."[12]

Like any new technology, EVs come with a learning curve. There will be mistakes, miscalculations, and missteps along the way. There will be hyperbole from EV supporters and detractors and from across the political spectrum.

This book is an attempt to sort fact from fiction, to examine the mistakes, as well as the amazing steps forward.

The EV story is not over.

Let us begin.

PART I

THE DEBATE

CARS, CARBONS, AND CLIMATE

CHAPTER 1

CLIMATE CHANGE WARS:

EV VERSUS O&G

It is arguably one of the most memorable scenes in American film history. Howard Beale, fictional TV anchor in the Academy Award–winning movie *Network*, learns that he's being let go, just minutes before what turns out to be his final evening broadcast. With cameras rolling, he devolves from unsettled to unhinged. His epic, on-camera meltdown ends with what became a semi-farcical mantra in the late 1970s:

> "I'm mad as hell and I'm not going to take this anymore!"

Some 40 years later, not only is that catchphrase still around, but it's also been adapted to modern times. Whatever anger was brewing in America late last century has percolated into full-blown fury. While Beale railed specifically against inflation, the Russians, and crime in the streets, in today's politically charged atmosphere, we're now outraged about nearly everything.

We are an agitated people.

It should surprise no one that the debate over climate change—*Is it real? Is it man-made? How bad is it? Who's to blame?*—has turned into a cultural touchstone. If you don't believe in the idea that climate change is man-made, well, off with your head! The Trump administration's decision to withdraw from the Paris Agreement further stoked the fire.

From coffee hour at the local doughnut shop to full-chamber debates on the Senate floor, discussions about climate change too often move quickly from polite conversation to full-throated cries: "My side is right and your side is stupid." We've gone from asking, "Why do you think that way?" to "How can you *possibly* think that way!" This reaction does more than shut down open, honest conversation. It feeds the ire and inspires left-right conflict and science motivated by politics instead of truth.

It wasn't always this way. Before the term "climate change" had even entered the public consciousness, presidential candidate George H. W. Bush, making a stump speech in Michigan, discussed the importance of treating land, water, and air with care, saying, "These issues know no ideology, no political boundaries. It's not a liberal or conservative thing we're talking about."[1]

What happened? How did environmental issues become politicized? And is there any way back?

Seeing the World Through Different Lenses

It's not easy to winnow out exactly how, why, or when climate change became an intractably divisive issue, but it does clearly cut across party lines.

An April 2018 Pew Research study indicated "wide political divides over the effects of climate policy." That includes how they view one of the key climate quarrels: whether oil and gas have a place in an ostensibly "zero-emission" future. For one thing, Pew found that 66 percent of liberals said policies aimed at reducing climate change are beneficial for the environment, compared with 27 percent of conservatives. On the flip side, conservatives see more risk to the economy: Fifty-seven percent responded that such policies are economically detrimental, while only 14 percent of liberals expressed these opinions.

But before they can agree on that topic, they have to reach détente on the science behind the debate. And that seems unlikely, especially given that "partisanship is a stronger factor in people's beliefs about climate change than is their level of knowledge and understanding about science."[2]

When the Green New Deal (GND), a campaign spearheaded by Democrats to eliminate carbon emissions, hit the congressional floor, the result was expectedly contentious. Senator Ed Markey (D-MA) and Representative Alexandria Ocasio-Cortez (D-NY) introduced a fourteen-page resolution in February 2019 calling for the United States to transition to 100 percent renewable, zero-emission energy sources within ten years. Support—and backlash—followed party lines: Congressional Democrats and liberal environmental groups jumped on the bandwagon, while President Donald Trump and other Republicans deemed it impossible at best. Mark Mills, senior fellow at the Manhattan Institute and a faculty fellow at Northwestern University's McCormick School of Engineering and Applied Science, went so far as to call the measures "immoral."[3]

Clearly, beliefs regarding the impact of fossil fuel emissions—something individual choices can influence through, say, driving habits—are disparate. Here, more than half of liberal Democrats say if more people drive hybrid and electric vehicles, it will make a difference. As for conservative Republicans, less than a quarter think it matters at all.

It doesn't take a research team to verify that maybe, just maybe, opinions on the value of electric versus gasoline-powered cars have something to do with how the respondent views America's oil and natural gas industry.

Granted, not every liberal shares fracktivist Bernie Sanders's vituperative desire to shut down the entire industry. One only has to read the balanced April 11, 2016, *New York Times* op-ed by Gary Sernovitz, a managing director of Lime Rock Management, to understand how a self-identified liberal can nonetheless appreciate the industry in general and fracking in particular. Sernovitz goes so far as to say that the

American shale revolution "advanced causes dear to most liberals' hearts"[4]—boosting employment, reducing poverty, keeping billions out of human rights–denying petro states, and reducing toxic greenhouse gases. In fact, because the natural gas unleashed by fracking can produce the same amount of electricity as coal with half the carbon dioxide emitted, coal has been increasingly displaced for use in electricity generation. The United States now leads the world in carbon-emissions reduction.

Even President Obama commended the contributions of domestic oil and gas production during his years in office. In his 2013 State of the Union address, for example, he said, in part, "After years of talking about it, we're finally poised to control our own energy future. We produce more oil at home than we have in fifteen years. We have doubled the distance our cars will go on a gallon of gas . . . [and] we produce more natural gas than ever before—and nearly everyone's energy bill is lower because of it. And over the last four years, our emissions of the dangerous carbon pollution that threatens our planet have actually fallen."[4]

Still, whether their beliefs are accurate or not, one need only look to the headlines to see that liberals often view the energy industry as backward, resistant to change, and more likely to cause environmental problems. The alternative energy industry? It's populated by altruists who have Earth's best interests at heart, they say. Republicans see oil and gas companies as an engine for economic growth and believe that abundant and affordable energy is the key to national security and prosperity.

The fact is the issue isn't black or white . . . but many shades of green.

Bulls, Beasts, and False Promises

Let's take a closer look at some of the assertions in support of electric vehicles (EVs), which are being hailed as the solution to everything from air quality—particularly in urban areas where "pollution belching vehicles poison the lungs of humans forced to share the streets with them"[5]—to economic growth.

New Jersey, a state more often associated with the smog over Newark than the verdant farmland worthy of the name Garden State, has an "awful lot to gain" by tapping into the EV market. At least that's the promise of an analysis presented at a December 2017 roundtable reported on—and sponsored by—NJ Spotlight.[6]

The research suggests that the emerging EV market in New Jersey could result in more than $2 billion in net economic benefits by 2035, even when the cost of installing the infrastructure and enhancing the electric power grid are taken into account. Mark Warner, whose company conducted the study, said that the widespread adoption of EVs will profoundly alter the electric-power market—and that's where the savings will accrue.

"EVs are going to flatten out the energy load of the entire grid, largely because most people will charge their cars at night, when the demand drops dramatically," Warner said. "When you change the load curve, you change the cost of energy."

In a piece for *Fortune* magazine, Joshua D. Rhodes and Michael E. Webber presented an even more bullish pro-EV position. They suggested that electrification of transportation would be nothing short of transformative—not only reducing pollution but also putting an end to wasted energy.

"The transportation industry consumes the most energy of all end use sectors, about 29 percent of total U.S. energy consumption, mostly in the form of petroleum products burned in internal combustion engines operating with about 25 percent efficiency," they wrote.[7]

The duo contended that, if every one of the *3 trillion miles* that automobiles drove over American roadways in 2017 were EVs that were 70 percent efficient, it would be equal to improving the efficiency of our entire economy from 32 percent to 34 percent. Sure, that two percent might seem negligible—but "a small change in a big number can make a huge difference."

Sounds pretty convincing, right?

Perhaps. But not everyone buys into the EV promise with the same level of enthusiasm or confidence.

Some critics complain that EVs are held up by subsidies—and they're not holding their weight on the environmental issue, either.

One problem is that the Energy Improvement and Extension Act of

2008, which granted tax credits for new qualified plug-in electric drive vehicles, gave what many perceive as an artificial boost to EV sales. As states repeal EV tax credit laws or let them expire—or even tack on fees that make electric cars more of an investment—the sense is that they haven't gotten their money's worth in terms of environmental benefits. With the incentives disappearing, consumers are turning away from EVs in droves: When Georgia scrapped its tax credit in 2015, EV sales fell from 1,300 cars in June to just 97 in August.[8]

And let's be clear: The EV tax credit was meant to stimulate innovation and sales, *not* to prop up the industry in perpetuity—which is precisely why the legislation built in a phase-out expiration that starts to wipe out the credit, in stages, once a model has sold 200,000 cars. In fact, because of their sales, Tesla and GM are facing some phase-outs. Tesla's Model 3 was down to $3,750 for sales between January and June 2019, dropped to $1,875 for the rest of 2019, and then will be $0. The Chevy Volt and Bolt phased down to $3,750 in April 2019, to $1,875 in October 2019, and then $0 in March 2020.[9]

Still, there's plenty of momentum, globally speaking.

Worldwide EV sales passed the 5 million mark in December 2018,[10] with plug-ins earning a 2.1 percent market share of new car sales.[11] This is an increase from 1.3 percent in 2017 and 0.86 percent in 2016.[12] Analysts at UBS predict that worldwide sales of EVs will reach 14.2 million units in less than a decade, which represents 13.7 percent of all new passenger cars.[13]

Will that projected growth be stymied by what some are calling the "dark side" of EVs—that they're only as clean as their power supply? Is the fact that it takes a lot of energy to produce the batteries and dispose of those batteries an environmental nightmare? And what about this: All over the world, a considerable amount of electricity is still generated by coal.

Those who think EVs have a long way to go maintain we shouldn't expect to see the end of fossil fuels anytime soon.

The oil and gas industry has already improved the efficiency of fuels. Shell Oil plans to spend up to $1 billion a year on alternative energy[14]— for comparison, its total 2018 capital expenditure (CAPEX) topped $23 billion[15]—and bought a charging station company, although it says efficiency gains in engine research translate to fuel savings that are three times those from the sale of EVs. Most large oil companies are investing heavily in natural gas in the belief that its relatively "clean" characteristics, compared with coal and oil, will keep it growing longer.

So which is it—are EVs an emissions panacea and savior of the planet, or are they a broken industry full of false promises held up by taxpayer subsidies? Is the oil and gas industry an aging beast that is determined to ruin our grandchildren's future, or a force for good?

And how do you make up your mind?

"Nothing Either Good or Bad"

In his book, *The Happiness Hypothesis, Finding Modern Truth in Ancient Wisdom*, author and professor of psychology Jonathan Haidt compared moral arguments—something like, "People who drive conventional cars are killing the planet"—with aesthetic judgment—"Gosh, that painting is beautiful."

Comparing concerns about fuel emissions with considerations of the Mona Lisa might seem a stretch. But bear with me.

Haidt contended that most people don't really know *why* they consider a piece of art beautiful. Like many complex answers, it can be boiled down to simple brain science. There's a part of the brain that makes up possible reasons why a painting appeals to you, and you go along with the first one that makes sense to you. Maybe it's the color, the symmetry, the wry smile—any number of things alone or in combination.

It's the same way, he said, when people are making a decision about important issues like climate change. Feelings come first, and then we go on a "cognitive mission" to find support for our beliefs or actions.

In other words, if you think EVs are the cure-all for the world's ills, you will look for research that agrees with you and discount—possibly even ignore—anything that doesn't. Conversely, if you believe that climate change isn't real or EVs don't hold up to the promises the industry makes, you'll accept only data that confirms your opinion.

The upshot, according to Haidt, is that we end up with "the illusion of objectivity. We really believe that our position is rationally and objectively justified."

It's not. We only hear what we want to hear. Or, as Haidt said, "There is nothing either good or bad, only thinking makes it so."[16]

BLINDED BY SCIENCE

But what about when we're looking for information that supports our side in the climate change war—or information that proves the other guy is wrong? If it's based on science, it must be objective, right?

Not so fast.

Adding to the quarrel over climate change is the fact that some of the science around it isn't very good. There are cases where science has been legislated only to discover it wasn't accurate to begin with.

Consider this: To meet mandatory targets for reducing greenhouse gases, some European nations promoted the use of diesel fuel for vehicles—and by "promoted" I mean provided generous financial lures to drivers (and, ultimately, car manufacturers). The theory was that because diesel-powered engines burn less fuel than gasoline-powered engines, they'd emit less carbon dioxide (CO_2). Germany, France, and nine other countries gave tax incentives on the purchase of diesel cars and fuel worth €21 billion ($25 billion) between 2014 and 2016. As a result, by 2017, 41 percent of all passenger cars were diesel-powered,[17] and, with 100 million diesel cars on the road, Europe had twice as many as the rest of the world altogether.[18]

But there's one problem—and it's a big one. Diesel engines might churn out less CO_2, but they far exceed gasoline engines in terms of spewing nitrogen oxides (NOx) into the air. In fact, European data released in January 2015 suggested that modern diesel cars produce ten times more toxic air pollution than heavy trucks and buses. NOx are the by-products of diesel most closely associated with harm to human health.[19] Damian Carrington wrote in a piece for *The Guardian* that NOx pollution is responsible for tens of thousands of early deaths across Europe, with the United Kingdom suffering a particularly high rate.[20]

In a sharp course reversal, some countries are now trying to end the sale of diesel cars. As *The New York Times* reported in August 2017, Madrid and Athens have banned them completely. Leave it to the Germans not to give up, however. Even after Volkswagen's "dieselgate"—where the company admitted equipping its cars with sensors that were used to cheat on emissions tests—Melissa Eddy and Jack Ewing from *The Times* noted that "politicians and automotive executives alike rejected calls by environmental groups to force carmakers to add antipollution hardware like better catalytic converters. And they were unanimous in opposing plans by some cities to ban diesels from downtown areas."[21]

CAN'T WE ALL JUST GET ALONG?

Aside from blinding people with (unintentionally) bad science, the sad truth is that, sometimes, the very people charged with providing impartial data are inserting their own opinions into the fray.

One example is the Union of Concerned Scientists.

Because they bill themselves as a nonprofit science advocacy group, it would be appropriate to think their mission is to promote the practical application of science—sharing important advances, encouraging science education, maybe getting into funding issues.

But visit their website and their bias is immediately apparent. Their fundraising appeal pits the organization and potential supporters against "notorious climate deniers" who have taken over "key science and environmental positions in Congress" and are pushing "ExxonMobil's anti-science agenda." For anyone who expects "scientists" to remain committed to an open-minded search for the truth, this sounds more like the words of a political action committee.

Sadly, they're not alone in trying to marshal sentiment for science as they see it. It doesn't take much more than a cursory glance at climatefeedback.org—which describes itself as a worldwide network of scientists who collectively assess the credibility of influential climate change media coverage—to see that the analyses they find credible are the ones that generally agree with the positions they support.

So much for the scientific method.

The important thing to remember here is that, really, we are all on the same side: No one wants to kill the planet. Everyone wants a good future for their grandchildren. By seeking accurate science, not just the science we want to hear, we can tone down the rhetoric, find a satisfying compromise, and maybe have that hand-holding "Kumbaya" moment after all.

CHAPTER 2

CARBON À LA CARTE:

THE TRANSPORTATION SMORGASBORD

You cannot separate the conversation of electric cars from the conversation of climate change. One need only look to the headlines to see that the topic of EVs is intrinsically woven into the conversation of carbon reduction. This conversation by extension is inseparably linked to the discussion about oil, energy, and the environment.

In many ways this association between EVs and climate is a shame. EVs are very cool cars in and of their own right—they are quieter to drive, they have speedy pickup . . . and wouldn't we all rather have our cars magically refilled at night via electricity, rather than breaking from our lives every 250 miles to stand and hold a smelly pump?

Unfortunately, just how cool EVs are is buried by the discussion of how *cooling* they are.

To attempt to address the issue of climate change takes us so far beyond the discussion of EVs that we would no longer be talking about cars. However, a point we can all agree on—outside of the climate debate—is that, in general, pollution is bad. Reducing automobile emissions is good. And reducing dependence on fossil fuels is good (if not imminent), not only for the environment, but for national security as well. A reduction in fossil fuel dependence protects against oil price shocks and reduces foreign imports.

EVs, of course, are only one of the many low- or zero-carbon transportation options being explored, for both the cool factor and the cool*ing* factor. A far cry from our choices just a few hundred years ago—"Should we take the horse or walk, dear?"—today's options for travel are legion. From golf carts to Segways, light rail to tram, taxis to ride sharing—and of course we can always choose the good old-fashioned horseless buggy. Moreover, cutting-edge transport marvels continue to multiply, zipping along exponentially, it seems.

If you're lucky enough to be living in one of the most advanced transportation ecosystems in the world—namely, Germany and shortly, Japan, China, and Canada—you might have the opportunity to hop onto an emissions-free mass transit system, something known as "hydrail" or hydrogen fuel cell-powered rail. The hydrail is an electric train that combines hydrogen with oxygen from the air inside a fuel cell, producing nothing but electricity for motive power along with steam and water—*no carbon dioxide at all.*

Hydrail was actually invented by Dr. Arnold Miller and the team at Golden, Colorado-based Vehicle Projects LLC, who built the first American hydrogen fuel cell-powered mining locomotive in 2002.[1] However, American railway operators have been reluctant to make any investment in a new kind of emission-less train.

Instead, we watched China introduce the world's first hydrogen trams in the city of Tangshen in October 2017.[2] Then, a year later, Germany put the first hydrogen trains into commercial passenger service in September 2018, running a 62-mile (100 km) route in the northern state of Lower Saxony[3]—a stretch of track that has been crowded with diesel trains for the last century. And other test zones around the globe are showing the growing viability of and interest in hydrogen-powered rail.

Americans do love a more private driving experience. And in keeping with our entrepreneurial cowboy spirit, perhaps it is no surprise that American futuristic transportation options have mostly come from companies like Uber and Lilium—firms trying to discover and stake their claim in the next frontier for individual city commuters, including autonomously piloted taxis and private vertical takeoff and landing (VTOL) vehicles.

Cultural Differences

Clearly, in the world's most advanced transportation economies, the choices for getting from here to there are increasingly complex. On close examination, though, it's obvious that global citizens make different choices based on their cultures, economic resources, and the transport infrastructures where they live and work.

Consider the Western Hemisphere. Until the 1960s, policies encouraging car travel in the United States served as a template for much of Western Europe. The U.S. auto industry was dominant, and gasoline prices extremely low. U.S. planners and engineers developed initial standards for roadways, bridges, tunnels, intersections, traffic signals, freeways, and car parking. To a lesser extent, Europe followed suit.

Then, in the late 1960s, European governments began curbing car use, especially in cities, by promoting walking, cycling, and public mass transit, primarily via train and tram. The objective was to make cities more livable and less smoggy.

In the United States, though, mass motorization became popular as the economy boomed. In Detroit, the automobile industry promoted the dream of big cars with very big gas-guzzling engines—the "fat cigar" or "land yacht" cars. Examples included the Oldsmobile Toronado (211 inches), Lincoln Continental (216 inches), Pontiac Bonneville (220 inches), Buick Electra 225 (originally 225 inches, elongated in later production years to 233.7 inches!), and Cadillac Fleetwood 75 (252 inches—a full 21 feet long).[4] The vehicles became symbols of the American good life—known for comfort, prestige, and poor fuel economy hovering around 11 mpg or less.

With the Arab Oil Crisis of 1973, followed by the second oil crisis of 1979, shocked Americans became aware of the vulnerability of gasoline supplies. For a brief time, smaller, more fuel-efficient cars became popular in the United States, though many models first introduced by American carmakers to emulate the more powerful European and Japanese economy cars were outright failures from both a mechanical and safety point of view. By 1975, however, the trend was clear: Small-vehicle imports from Japan and Germany had shot up to more than 15 percent of U.S. vehicle totals. That same year Congress enacted the Corporate Average Fuel Economy (CAFE) standards, mandating 27.5 miles per gallon within ten years.[5] CAFE in fact became reality for most car manufacturers by 1985; the fuel standard also languished for the next twenty-six years until Congress acted again in 2011 to implement tougher fuel economy regulations.

Today, the United States has two parallel sets of fuel standards, both CAFE, administered by the National Highway Traffic Safety Administration (NHTSA), and the EPA-controlled greenhouse gas (GHG) emissions standards. In 2012, the government harmonized both sets of standards to require much greater fuel economy and lower carbon dioxide emissions. By 2025, for example, CAFE would have required light-vehicle fleets to maintain fuel economy between 48.7 and 49.7 mpg (before CAFE credits and flexibilities).

As for GHG emissions standards, the EPA stipulated a fleet-wide level of 163 grams per mile of CO_2 in model year 2025, which is equivalent to 54.5 mpg (if achieved exclusively through fuel economy improvements). The overall goal was to cut 6 million metric tons of greenhouse gas over the lifetime of U.S. vehicles sold in model years 2012 through 2025, and to save American families $1.7 million in fuel costs.[6] Automakers that failed to meet CAFE targets would pay noncompliance fines after a lead-in transition period. However, the GHG emissions standards cannot be manipulated or shirked by automakers paying noncompliance fees.

These rulings strongly suggested that U.S. standards organizations were serious about carbon dioxide emissions; the government's stated goal is to reduce U.S. dependence on oil by 2 million barrels per day by 2025.[7] In addition, the EPA is also targeting reduced nitrous oxide and methane emissions. These efforts include extending a number of carbon dioxide credit programs to encourage continuous car manufacturer improvements of vehicle air-conditioning (i.e., use of alternative refrigerants or lower refrigerant leakage), as well as GHG and fuel economy technologies. These low-carbon alternatives are the kinds implemented in the new generation of EVs, plug-in hybrids, and fuel cell vehicles—all of which are becoming increasingly noticed, if not popular.

In March 2020, however, the Trump administration cut the year-over-year improvements expected from the auto industry, slashing standards that require automakers to produce fleets averaging nearly 55 mpg by 2025 down to about 40 mpg by 2026. The Hill reported that the new regulations bring mileage below what even automakers said is possible for them to achieve. Moreover, in April 2020, NHTSA and EPA announced amended CAFE and GHG emissions standards for passenger cars and light trucks and established new less stringent standards, covering model years 2021 through 2026.

Our Carbon Reduction Options: Changing Behaviors, Not Just Technology

But is the average American consumer actively looking toward alternative transport? The short answer? Not really. A perfect example is the anachronism known as the American passenger train. Except for certain profitable and much-traveled routes in the Northeast (e.g., New York, Pennsylvania, Boston, and Washington, D.C.), the federally owned and operated national train company, Amtrak, has encountered continuous, even tragically predictable financial and service troubles throughout its nearly five-decade history.

Founded in 1971, Amtrak operates 44 routes on 21,000 miles of track in 46 states.[8] Because freight rail companies own about 95 percent of the track (Amtrak owns its own trains), Amtrak is subject to high leasing fees and maintenance agreements. Analysts decry federal micromanagement and a seeming inability of the train corporation to respond nimbly to market conditions.

Amtrak has a record of losing money on most of its routes and it has been criticized for having a "weak safety culture."[9] In 2018, Amtrak had total revenues of $3.3 billion and total operating expenses of $4.2 billion, receiving direct federal subsidies of $1.9 billion.[10] In November 2019, however, Amtrak executives announced that in the fiscal year just ended, the service inched closer to breaking even and reported record growth on the heavily traveled Northeast corridor and state supporter lines. Repeated calls to privatize Amtrak have fallen on deaf ears, even though privatization of comparable passenger rail systems in Britain, Japan, and Canada has increased ridership and revenues while decreasing fares and cutting operating costs.

But in America, the overwhelming preference of consumers is the private car, even with the negatives of traffic jams, suburban sprawl, and long-distance commutes. According to Ralph Buehler, an analyst of city transportation trends, "In 2010, 85 percent of daily trips in America were taken in cars, compared to 50 to 65 percent in Europe. Trip distance partially explains the discrepancy, but while about 30 percent of trips were less than a mile, Americans drove the distance 70 percent of the time, while Europeans chose other means, including riding, biking and public transportation."[11]

Today, the Europeans are starting to catch up to the Americans in car usage. Eurostat found that, as of April 2018, "the passenger car fleet in almost all of the EU Member States has grown over the last five years."[12] Passenger travel in private vehicles now constitutes roughly 83.7 percent of all European trips. Between 2004 and 2014, passenger train travel increased somewhat, from 6.7 percent to 7.6 percent, while the use of motor coaches, buses, and trolley buses decreased from 9.9 percent to 9.1.[13]

In the United States, awareness and carbon pollution has principally been met by improvements in technologies, not behaviors. Car companies continuously try for better fuel economy and less polluting air-conditioning and fuel alternatives. Comparatively little investment is made in encouraging car users to get off the roads, despite the publicity and excitement when five U.S. cities—Buffalo, Portland, Sacramento,

San Diego, and San Jose—launched their light rail systems in the 1980s. The results of these light rail systems have been mixed.

Although billions in federal, state, and local dollars have been spent to build 650 miles of light rail lines in sixteen regions since that period—and today more than 144 miles of additional lines are under construction—no region has invested in a new heavy rail subsystem since 1993. Moreover, a quick examination of ridership shows that the initial five light rail metro systems failed to resuscitate city-center living, nor did they produce the higher transit use city planners had hoped for. These very same cities that were pumped about their trains also built major toll-free highways during the same period to accommodate sprawling, diverse car traffic.

Each metro area also continued to sprawl, and, most notably, local government and transit planners failed to address the comparatively low costs of commuters continuing to drive and park. As a result, light rail produced only mediocre results, and the American city and suburban dwellers continued to choose traffic in private cars rather than switching in large numbers to trains and trams.[14]

Sharing a Ride—or Skipping It Altogether

Rail may not be taking off as a popular American alternative to the private car. But a few transportation-related behavioral changes appear to be gaining traction in American metro areas.

Consider the uptick in the popularity of ride sharing. According to one study, it is projected that between 2018 and 2022 ride sharing—as in sharing private cars and getting to places with drivers you don't necessarily know—is to experience an annual growth rate of 15.9 percent, which will mean a market volume of $28.371 million in 2022.[15] What's more, "transport as service" companies—think Uber, Lyft, and the like—are sprouting up in virtually every country on earth.

The big incentive for ride sharing is the lower fees that make these services a more economical alternative to conventional city taxis. Another plus? Nearly all ride-sharing and carpooling services are automated and easily accessible through smartphone apps, GPS, and social networks.

In Africa, for example, where rapid urbanization is way ahead of infrastructure build-outs, real-time ride sharing is disrupting the transportation industry. It's estimated that almost 40 percent of Africans live in urban areas, a percentage expected to increase to 50 percent by 2030 and 60 percent by 2050. Ride sharing—Africa now has 56 different ride-sharing services—is now considered one of the best and most economical alternatives to car ownership. The practice not only offsets the costs of owning a vehicle, but it also reduces the impact of gasoline shortages while drastically reducing the amount of CO_2 emitted by cars. Presumably, it helps professionals in big cities to beat traffic.[16]

But Uber isn't the only way today's professionals can beat traffic— and reduce their need for transportation, shared or otherwise. According to the research company Global Workplace Analytics, 50 percent of the American workforce can benefit from teleworking at home at least part-time, while 80 to 90 percent of the workforce said they would like to telecommute—the most highly popular option is two to three days a week. Moreover, telecommuting has increased 115 percent for regular U.S. employees (not self-employed) since 2005. Forty percent more U.S. employers today offer teleworking options than five years ago. More than 3.7 million employees (2.8 percent of the total workforce) are actively teleworking at least half the time. Larger companies are more likely to offer flexible options to employees than the smaller ones, and the highest concentration of teleworking happens to be in the busiest regional corridors—mostly New England and the mid-Atlantic states.

The typical profile of a telecommuter is college educated, 45 years old and older, earning an annual salary of $58,000 or more. Seventy-five percent of those who do work at home earn more than $65,000 per year, placing them in the top 80th percentile for home or office workers. The potential for knowledge workers, especially those in Fortune 1000 companies, to remain mobile or telework is enormous; already, these large companies are revamping their office space to accommodate employees who are not at their desks 50 to 60 percent of the time.[17]

Combinations of work and mobility will become even more commonplace as transportation technologies evolve throughout the 21st century.

All Flavors of Electrification

And what of electrification? When we refer to an "electric car," it's natural to assume we are talking about a single type of car. In reality, however, electrics come in many flavors, each with varying levels of carbon savings.

A pure electric vehicle (EV), also known as a BEV (battery-electric vehicle), uses batteries exclusively to power the car; there is no hybrid or gas-electric engine to supply the battery. An "electrified" car, by contrast, means that numerous components of the car receive electricity to operate. An example of "electrification" is a plug-in hybrid gas-electric vehicle, which combines both battery operation and a gasoline engine to produce the power to drive the car.

In the past, a car was powered exclusively by internal combustion—a gasoline-fueled engine delivering mechanical force to axles and wheels (a separate battery was used for engine starts and to power such functions as heat, lights, radio, and air-conditioning). Today, EVs are varied in the extent of their electrification; the category of car is dependent on the source and distribution of electric power along with the primary method(s) of achieving locomotion. Arguably, there are a few major categories of vehicles: electric, alternative fuel (hydrogen, biofuel/ethanol), and fossil fuel (including gasoline and diesel). So let's delve into each category (and subcategory), along with respective benefits and drawbacks of each.

ELECTRIC VEHICLES (EVS) OR BATTERY-ELECTRIC VEHICLES (BEVS)

These vehicles are gasoline-free, generally using a lithium-ion battery to store the electric energy that powers the motor. These vehicles need to be plugged into a charging station or an ordinary electrical outlet to recharge. EVs offer reduced carbon emissions, although the charging source for an electric car may be a "dirty" electric grid (i.e., a grid powered by fossil fuels such as coal or natural gas). Acceptance of pure EVs has been growing steadily since the introduction of such vehicles as the $30,000 Nissan Leaf, and the luxury $80,000 Tesla Model S. However, "range anxiety" continues to be a deterrent to potential buyers. As the number of charging stations outside major cities grows, addressing that concern, the number of EVs on the road are expected to increase rapidly.[18]

Hybrid Electric Vehicles (HEVs)

With an internal combustion engine that primarily runs on fuel, a hybrid also uses an electric motor powered by electricity stored in a battery. A gas-electric hybrid is considered a stable, albeit transitional, technology offering the benefits of gasoline reliability and range along with lower emissions.[19] The Toyota Prius is a classic example. Its battery charges via regenerative braking and the internal combustion engine; therefore, the car does not require a "plug in" to a standard electrical outlet or charging station in order to run.

Plug-in Hybrid Electric Vehicles (PHEVs)

While running primarily on fuel and an internal combustion engine, this vehicle also uses an electric motor and battery-stored energy. Recharging is accomplished by plugging into an electric power source, which can provide more than seventy miles of driving distance for some. Running solely on gasoline, which is similar to a conventional gas-electric hybrid, is also an option with these vehicles. A classic PHEV is the 2018 Chevy Volt, offering 53 pure electric miles and up to 420 miles with a full charge and a full tank of gas. (GM discontinued production of the Volt in 2019, citing lagging sales and consumer preferences for SUVs.)

ALTERNATIVE FUEL VEHICLES

For those who like a little variety in their transport menu, individual and fleet vehicle owners may pick from a buffet of alternative fuel vehicles of the nonelectric kind.

Flex Fuel and Biofuel Vehicles

Today, more than 80 cars and trucks are marketed as "flex fuel" vehicles, which are powered by a combination of gasoline and ethanol, with ethanol making up the larger portion (gas-ethanol blends can contain 85 percent of the corn-derived fuel). Although ethanol is considered an environmentally friendly "renewable" energy source, some critics are concerned that leaning too heavily on "flex fuels" will drive corn prices up or take a bite out of corn that is typically grown for food.

Biodiesel Vehicles

Biodiesels raise the octane rating of conventional diesel fuel, increase lubricity, and burn more cleanly; generally, biodiesel fuels are mixed with conventional sources and can be used in conventional petroleum diesel engines without modification. These vehicles accept up to 20 percent of diesel fuel from sources such as vegetable oil, animal fats, and recycled restaurant grease.

Propane Vehicles

Also called liquefied petroleum gas (LPG), propane is a by-product created during the processing of natural gas and refining of crude oil. Propane is a common alternative fuel for light-vehicle fleets (e.g., buses and police vehicles) along with heavy-duty trucks. With more than 270,000 of these estimated on the road in the United States today,[20] propane vehicles offer reduced emissions and comparatively easy maintenance.

Liquefied and Compressed Natural Gas (CNG) Vehicles

Vehicles using either liquefied or compressed natural gas burn cleaner than gasoline and get about the same fuel mileage. More than 112,000 vehicles in the United States, most of them heavy- or medium-duty trucks, are powered by natural gas. A CNG vehicle is generally slower than a gasoline one and offers limited range and refueling options. Trucks powered by natural gas also cost more but offer advantages of reduced fuel costs and emissions.[21]

Hydrogen Fuel Cell Vehicles (FCVs)

As both a combustion additive and an energy source for generating electricity, hydrogen processed in a fuel cell will combine with oxygen, emitting nothing but electrons for power and water at the tailpipe.[22] FCVs emit no carbon dioxide and are said to be two or three times more efficient than gasoline-powered vehicles.

However, fuel cells are expensive to build, along with a network of fueling stations to distribute the hydrogen. Early models, the Honda FCX Clarity and the Mercedes-Benz F-cell, both have been discontinued. About 8,000 hydrogen fuel cell cars have been sold in the United States since 2012. After a basically stagnant year in 2018, 2019 sales showed a decline of 12 percent, according to the California Fuel Cell Partnership. Scientists need to develop better ways for fuel cells to store and distribute the hydrogen in ways that can be scaled to support a large transport system.

FOSSIL FUEL VEHICLES

Traditional fossil fuel vehicles have been optimized for unprecedented levels of efficiency. The advent of Mazda's new homogeneous charge compression ignition (HCCI) technology promises to make gasoline cars 25 to 30 percent more efficient than they are today.

Human Factors Prevail

Within the next five years, American urban commuters will begin to see the development and launch of air taxis and flying vehicles that will be piloted autonomously, by remote control, or by pilots. As many as 15 startup companies, along with established players (e.g., Airbus) and private electric aircraft ventures are in development today, perfecting alternative vehicles that move traffic from the roads to the air.

As urban vertical takeoff vehicles and taxis take hold, much the way drones and surveillance satellites work currently, the entire infrastructure and pattern of intercity commuting will very likely change.

Even today, American real estate developers and construction companies are touting "mixed use" commercial and residential communities to co-locate both workplace and living quarters for residents, in effect reducing the need for conventional car transportation and commuting.

But even with these trends, the future of transportation remains up for grabs. While we can't entirely predict traffic and population patterns, we can be assured that almost every form of transportation that survives will be cleaner and more fuel efficient than in the past.

This is not just the legacy of climate change. It's the desire for continuous improvements—less cost, better options, and ultimately, more convenient ways to live, travel, and work.

CHAPTER 3

ENERGY IN/ ENERGY OUT:

EV'S CARBON FOOTPRINT

You've likely heard the saying, "There ain't no such thing as a free lunch."

In the 1880s, American and European saloons would offer working-class, price-conscious patrons a "free" lunch if they purchased a single beer. This was, as we'd say today, a brilliant marketing move: A typical saloon spread might include liverwurst, pigs' feet, bologna, cheese, and salt-encrusted pretzels.[1] Although statistics are hard to come by, it's safe to assume that most nineteenth-century prospectors and factory workers bought two, three, or four more beers to wash down all the cheese, pigs' feet, and pretzels.

Popularized in Robert A. Heinlein's sci-fi Libertarian classic *The Moon Is a Harsh Mistress*, this phrase has become shorthand for opportunity cost and has been applied to everything from technology to sports to economics to the second law of thermodynamics.

It also applies to EVs. Although they can offer some environmental benefits, there is no such thing as an "emissions-free" car. When marketing EVs to energy-conscious consumers, companies often leave out some of the details—the bologna and pigs' feet, if you will, that make EVs costlier than they seem.

Electric cars on the road rely on regular charging from the electric grid, and the power plants providing that energy are not emissions-free. Even in energy-conscious California, where the state's electricity in March 2017 briefly topped 50 percent from solar sources,[2] most remaining power in the grid has derived from conventional fossil fuels generating emissions. Consider the 2017 California energy mix: Natural gas supplied 33.67 percent of the state's electricity, and renewables (biomass, geothermal, small hydro, solar, and wind) made up 29 percent of the state's electricity, so natural gas still claimed the top spot.[3]

Germany, by contrast, claims 85 percent to 90 percent in renewable electrical power on sunny or windy days. China, on the other hand, the biggest greenhouse gas polluter in the world, has instituted measures to roll back its power plant consumption of coal and increase its share of nuclear and renewables.[4] However, coal still generated 72 percent of all Chinese power in 2014, according to the International Energy Agency (IEA), and more recent Chinese government reports stated that the Middle Kingdom is burning far more coal annually than previously thought—about 17 percent more than initially reported.[5] That's *almost a billion more tons of carbon dioxide* released into the atmosphere annually.

Discouraged? Consider cases closer to home. It's better news, but not perfect.

Straight from the Source, Power Matters

Back in the United States, old-fashioned coal continues to supply roughly 30 to 33 percent of all power needs despite an Obama-era move to retire coal-fired electric for less polluting natural gas power generation.

Coal, whether "soft" or hard—in other words, bituminous, subbituminous, or anthracite—contains sulfur. When burned, the toxic output of sulfur in the atmosphere produces acid rain. In addition, coal combustion generates toxic substances such as nitrogen oxides (key contributors to ozone and respiratory illnesses), soot (associated with chronic bronchitis and aggravated asthma), and mercury (a neurotoxin contaminating fish and wildlife, even causing birth defects). Carbon monoxide, volatile organic compounds, arsenic, lead, cadmium, and other toxic heavy metals are also expelled into the atmosphere.

After coal is incinerated, the remaining ash and sludge are often deposited in unlined landfills and reservoirs; this waste can easily contaminate drinking water and harm local ecosystems, as public outrage over the 2014 Dan River coal ash accident in North Carolina illustrated. In all, coal-fired electricity produced in America's utilities still accounts for about 24 percent of all energy-related carbon dioxide emissions today.[6] In addition, coal transport to and from states constitutes about one-third of *all* U.S. freight train traffic. These trains, as well as the trucks and barges, still run on diesel—a major source of toxic chemicals emitted into the air, including nitrogen oxide and soot.

There are 33 impoundments holding more than 100 million tons of coal ash in 14 locations around North Carolina. None of these coal ash reservoirs is lined, and, like the current and former power plants that generated the ash, nearly every one is within a few hundred yards of a major river or tributary.

The most natural starting point for the contemporary discussion of coal ash arose in North Carolina on Sunday, Feb. 2, 2014. A stormwater pipe collapsed and started to drain 39,000 tons of coal ash into the Dan River near Eden, North Carolina. But the 2014 spill was by no means the beginning of the story. Questions about what to do with the millions of tons of coal ash building up in basins across North Carolina started long before. In the decade prior to Dan River, there were a handful of incidents around the state, including smaller spills, leaks, and seeps, and dams overtopped in floods. A major spill in Tennessee in 2008 had also raised public awareness.

Given the rather lengthy list of coal's drawbacks and low natural gas prices, it shouldn't surprise anyone that natural gas has emerged as the new hero of U.S. power plant generation. After all, natural gas, a fossil fuel, emits the same amount of energy but 45 to 60 percent less CO_2 and fewer air pollutants than coal-fired plants when burned in the newer energy-efficient plants. According to the U.S. Energy Information Association (EIA), about 117 pounds of carbon dioxide are produced from combustion of natural gas per million British thermal units (MMBTUs). Compare that to more than 200 pounds of carbon dioxide per MMBTU of coal and 160 pounds per MMBTU of distillate fuel oil.

Natural gas–fired plants are certainly good news for EV drivers charging their cars in states where coal is no longer king. However, this clean-burning energy "lunch" still isn't free. When methane (CH4, the main component of natural gas) leaks from pipes or wells during digging and extraction, the gas is 34 times as potent as CO_2 in trapping heat in the atmosphere; that calculation holds true over a hundred-year period. Methane's heat-trapping capacity is even higher—86 times as high as carbon dioxide when measured over a two-decade period.[7]

The good news is that the oil and gas industry has been clamping down to contain methane leaks. A 2015 Washington State study monitoring 13 major gas distribution centers around the country found that methane emissions have been reduced by 36 percent to 70 percent compared to estimates published by the Environmental Protection Agency (EPA) in 2011.[8] A large part of the improvement comes from equipment and upgrades of leaky cast-iron pipes. One research study in 2012 found that over a 20-year period, methane leaks in power plants must be contained at less than 3.2 percent of total volume of gas processed to have lower life cycle emissions than new coal-fired plants. Preferably, the study continued, methane losses should be kept below 1 percent and 1.6 percent compared with diesel fuel and gasoline, respectively, to offer true environmental benefits.[9]

What's the bottom line? If you're charging your electric car in a state that relies heavily on coal for power generation, your "emission-less" EV is still generating significant amounts of greenhouse gas and other toxic air pollutants. If you're on a natural gas power grid, the plant is still generating nitrogen oxides, a precursor to smog, but at lower volumes than gasoline and diesel used for conventional motor vehicles.

That's great news. But it's not a free lunch. Some energy experts argue that the EV industry has simply shifted much (though not all) of the burden of air pollution on the road back to the power plant. Therefore, be cognizant that the cleanliness of an EV depends not just on what comes out of a tailpipe, but also on the power-generating sources back at the grid. Further, when you add energy expended in manufacturing an electric versus a fossil fuel car, and then consider the emissions picture over the lifetime, the picture becomes even more complicated.

The Manufacturing Equation

First things first; let's give credit where credit is due. EVs, overall, do have a smaller carbon footprint than gasoline-powered cars if you count driving in the long term—that is, over the average life of the vehicle.[10]

According to the U.S. Department of Energy Alternative Fuel's Data Center, "Hybrid electric vehicles (HEVs), plug-in hybrid electric vehicles (PHEVs), and all-electric vehicles (EVs) typically produce lower tailpipe emissions than conventional vehicles do."[11] But the benefit is variable by state and power supply.

"EVs and PHEVs running only on electricity have zero tailpipe emissions, but emissions may be produced by the source of electrical power, such as a power plant," the Alternative Fuel's Data Center's datasheet reads. "In geographic areas that use relatively low-polluting energy sources for electricity generation, PHEVs and EVs typically have a well-to-wheel emissions advantage over similar conventional vehicles running on gasoline or diesel. In regions that depend heavily on conventional fossil fuels for electricity generation, PHEVs may not demonstrate a well-to-wheel emissions benefit."

Despite regional variances, there seems to be general agreement that, on average, EVs are cleaner to *drive* than ICEs. That being said, they are very dirty to *make*. During the manufacturing process, an electric car produces *more* carbon dioxide—*not* less—and requires significantly more energy to assemble than a conventional fossil fuel car. Much of the discrepancy derives from the mining and manufacturing toll of the energy-dense lithium-ion battery. Mercedes, for example, reports that manufacturing a B-Class EV produces 45 percent of the car's total emissions, while a conventional B-Class Mercedes emits only 18 percent on the factory floor.[12] And a 2017, controversial IVL Swedish Environmental Research Institute study further claimed that big electric vehicles like Tesla, with its 100 kilowatt-hour (kWh) battery, can generate remarkably high levels of CO_2 emissions on the factory floor—as much as 8.2 years' worth of conventional driving in a gasoline-powered car.[13]

The Swedish study also found that for every kilowatt-hour of storage capacity built into a lithium-ion battery, emissions of 330 to 440 pounds of carbon dioxide will result just from manufacturing. Mining and extraction of rare earths, including lithium and toxic cobalt, ironically, only account for about 10 to 20 percent of the undesirable emissions according to the Swedes. Where did the other 80 percent of the emissions come from? The processing and production of lithium-ion batteries.

The Swedish researchers didn't evaluate batteries by individual brand. At the same time, their study also made several inaccurate assumptions, among them that gasoline and diesel fuels have no associated carbon footprint before they arrive at the fueling station. (They do; drilling, extraction, refinement, and transport also release greenhouse gases.) The study further assumed that Tesla vehicles are manufactured using 50 percent fossil fuel–generated power. (They are not; Teslas, built in California, are assembled from 100 percent renewable energy sources.[14])

But even with these inaccuracies, the Swedish study did provoke the industry with one telling recommendation: To save emissions, EV manufacturers would do well to optimize smaller lithium-ion batteries rather than zooming up to the biggest and most powerful batteries just to please the luxury market (target: Tesla). The researchers cited power requirements for a Nissan Leaf versus the Tesla Model S—batteries generating 30 kilowatt-hours and 100 kilowatt-hours respectively—claiming each type of battery emits 5.3 tons and 17.5 tons of CO_2, respectively, even before the cars hit the showroom for purchase. (For purposes of comparison, a Stockholm–to–New York round trip by air produces a little more than a half ton of carbon dioxide per person.[15])

Though many reputable automobile authorities rushed to dispute these numbers—*Popular Mechanics*, for instance, diligently recalculated the emissions of Tesla's batteries, comparing them favorably with fossil fuel–powered cars, and then pronounced the study "bunk"—the Swedish research also poked at a few sacred cows, reminding the media to research carefully the true "well-to-wheel" costs of any car option.

In the Tesla Model S case, *Popular Mechanics* author Ezra Dyer compared it to the gas-powered Audi A8 4.0: "According to the EPA, that car emits 6.2 metric tons of CO_2 per year, given 15,000 miles of annual driving. And since A8s don't automatically percolate their own 93-octane, the EPA also calculates an additional 1.1 tons of upstream carbon to get those ancient dinosaur innards coursing through your fuel pump. Math aficionados will note that 17.5 (battery production [of emissions]) divided by 7.3 (total annual A8 emissions) equals 2.4. As in, apples to apples, the battery's carbon footprint is zeroed out in less than three years."[16]

Three years of driving is still a high number to "zero out" the Tesla's initial excess of carbon dioxide pollution.

In addition, how much we save in emissions depends on how much we drive—and how long the car lasts.

The average life of EVs versus internal combustion vehicles varies according to estimates, and with ongoing technology improvements for both kinds of vehicles, no one knows exactly how long the newest engineering miracles will last. Some experts claim current EVs can operate for up to 400,000 miles because the power train has fewer parts and is easier to maintain than conventional cars; moreover, lithium-ion batteries are replaceable. The average life of an internal combustion vehicle, by contrast, is frequently estimated at around 200,000 miles or about eleven to twelve years of average driving, although new breakthroughs in internal combustion efficiency could lengthen average driving life considerably.

The longer an EV is on the road, the more carbon efficient it becomes. On the other hand, if the owner rarely drives, their EV purchase may have done more harm than good, emissions-wise.

The conclusion? It's certainly fair to say an electric car produces higher greenhouse gas emissions in the mining and manufacturing stage. This is especially true when you take into account the emissions released during the manufacturing of lithium-ion batteries, a process that can produce emissions of 330 to 440 pounds of carbon dioxide (this does not include the emissions associated with rare earth mining, another significant source of emissions and pollution).

But after an EV is purchased and the driver leaves a dealer's lot, conventional fossil fuel cars eventually catch up and pass the electric car in total lifetime emissions. In fact, one gallon of gasoline consumes the same energy as 20 miles of driving in a Model S. According to Department of Energy data, burning one gallon of conventional gasoline produces about 22.4 pounds of CO_2 emitted from the fossil fuel content, and burning one gallon of E20 (which combines gasoline with about 20 percent biodiesel) produces about 17.6 pounds of CO_2.[17]

GETTING 20 POUNDS OF CARBON
FROM 6 POUNDS OF GASOLINE

How can a gallon of gasoline weighing about 6.3 pounds suddenly produce 20 pounds of CO_2 when burned? The answer is that most of the weight doesn't come from the gasoline itself, but from the addition of oxygen molecules in the air during combustion.[18]

For context, it's interesting to note how much carbon conventional combustion engines emit. In December 2018, the U.S. Energy Information Administration estimated that U.S. motor gasoline and diesel fuel consumption for transportation produced 1,549 million metric tons of CO_2 (1,102 million metric tons via gas and 437 million metric tons via diesel, respectively). This figure equals 81 percent of the total U.S. transportation sector CO_2 emissions and 30 percent of the total U.S. energy-related CO_2 emissions in 2017.

All expenditures considered, including the energy and emissions costs of getting gasoline to the pump, CO_2 emissions can be four times less per mile for electric versus gas-powered vehicles. Consequently, while the manufacturing emissions argument is tantalizing for internal combustion advocates, the average numbers still come out in favor of EVs.[19]

Of course, the conversation doesn't stop there.

Issues Unresolved

Perhaps unsurprisingly, there are several other important aspects of the EV versus ICE debate that need to be addressed. One of them is the rapid upgrades and innovations in gasoline combustion technologies that have impacts on efficiency. In recent years, for example, technical improvements like direct fuel injection, variable valve timing, and cylinder shutdown systems, along with the addition of lightweight body materials and dual-clutch transmissions, have improved mileage per gallon considerably.

In August 2017, Mazda announced the rollout of the Skyactiv-X engine, introducing a combustion method commonly known as *homogeneous charge compression ignition*, which produces 20 to 30 percent greater energy efficiency than the company's best existing engines.

"Researchers around the world have tried to crack this process for years, but it has never really left the laboratory," reported *The New York Times* in an August 2017 article about Mazda titled "The Internal Combustion Engine Is Not Dead Yet." In the Skyactiv-X engine, combustion happens using extremely lean mixtures combined with air and ignited without spark plugs, much like a diesel engine. Praising the improvements, *New York Times* author Norman Mayersohn pronounced the Mazda combustion technology "the latest plot twist in a century of improvements for internal combustion engines, a power source pronounced dead many times that has persisted nevertheless."[20]

Mazda is set to release its Mazda3 sedan with the Skyactiv-X engine, which could very easily skew the EV-internal combustion debate yet again. But we're not holding our breath: In late April 2019, a company spokesperson said, "Skyactiv-X will be available in late 2019 in select markets as part of a phased rollout. Timing of U.S. availability has not yet been announced."[21] (In January 2020, *Road & Track* reported that, after arriving in dealerships in Europe and Japan in 2019, the car still has not arrived in the United States. A Mazda executive expressed doubt that the Skyactiv-X would appeal to American buyers, who he said value power over efficiency.) Mayersohn cited the predictions of John Heywood, a professor of mechanical engineering at the Massachusetts Institute of Technology, who believes that hybrid gas-electric or diesel-electric solutions will become more popular thanks to these improvements. Heywood also believes that EVs powered purely by batteries will comprise only about 15 percent of the vehicles sold by 2050. He estimates that 60 percent of light-duty vehicles will still have combustion engines at that time, some of them gas-electric hybrid systems equipped with turbochargers to maximize power (and minimize friction) in smaller engines. Heywood believes conventional fuel economy could easily double in 30 years. A quarter to a third of that improvement would come from areas like aerodynamics and weight reduction in the vehicle.

Ending the Confusion: Back to Power Plant Emissions

Comparing emissions figures can be perplexing. For example, the all-electric Nissan Leaf produces about the same amount of greenhouse gas pollution per mile as the hybrid Toyota Prius—even though the Leaf never burns gasoline and the Prius uses a combination of gasoline and electric batteries.[22] Adding to the confusion, researchers have found that electric plugs and sockets are not always equally efficient, with some losing electricity in the charging process.

But certainly more troubling is that EV manufacturing emissions can be 45 percent of the vehicle's lifetime total[23] (although 15 to 26 percent is a more standard range[24]). And because lithium-ion and other rare earths (cobalt, cadmium, chromium, mercury, and lead—all groundwater contaminants) are sourced from faraway places, mostly China, they add to the pollution and cost of EVs, posing a threat to public health, especially in exposed children.[25]

No free lunch, remember?

Clearly, our dilemma isn't solved. And we might do well to consider subsea hydraulics engineer Ryan Carlyle's conclusion drawn from a series of calculations about the best ways to reduce our carbon footprint. If we were to switch over to almost 100 percent EVs on the road today, we would save 417 million in carbon. To put that into perspective, in 2012, the total CO_2 emitted in the United States was 6.5 billion tons. That makes our emissions savings a somewhat disheartening 6.4 percent. Given the comparatively modest savings in carbon dioxide levels for the transition costs, Carlyle believes EVs are "a pretty crummy way to reduce CO_2 emissions, given the current U.S. power mix."

"You can do three times as much *good per dollar* by fitting coal plants with carbon capture systems," he continued. "Mass rollout of electric vehicles is only worthwhile in tandem with massive increases in renewables generation. Perhaps in the future we'll get there. But today's generation market trends do not support that assumption for the next several decades."[26]

This raises an important question: EVs (at least for now) almost always produce less carbon emissions than ICEs, despite high manufacturing emissions. But does that necessarily mean that EVs are a good way to reduce carbon?

The answer may not be as simple as we'd expect.

CHAPTER 4

IT'S COMPLICATED:

CLIMATE CHANGE, CO$_2$, AND COST

In early 2017, the U.S. Fish and Wildlife Service (FWS) made national headlines with a bombshell announcement: The West Indian manatee, which had been classified as an endangered species since the 1970s, was officially bumped down to the "threatened" list. In a press release issued on March 30—just one day after Manatee Appreciation Day—U.S. Secretary of the Interior Ryan Zinke called the moment a "milestone" that wouldn't have been possible without buy-in from multiple organizations.

"The Fish and Wildlife Service has worked hand in hand with state and local governments, businesses, industry, and countless stakeholders over many years to protect and restore a mammal that is cherished by people around the world," he said. "Without this type of collaboration and the commitment of state and local partners, this downlisting would not have been possible."[1]

For nearly four decades, hundreds of West Indian manatees have been killed or gravely injured because of encounters with (mostly oblivious) humans. The aquatic mammals are especially at risk during the winter months, when they migrate en masse to the Florida coast in search of warmer waters. Unfortunately, this also means that they spend half the year living in perilous proximity to humans, who are drawn to the coast in search of balmy temperatures and places to launch their speedboats and jet skis.

It could have been a recipe for extinction—and it almost was. In 1991, the manatee population in Florida had dwindled to a mere 1,267. In 2016, that number had climbed to just over 6,000.[2]

What do manatees have to do with EVs? Well, nothing. What's interesting—and relevant—about this example is the way the problem was solved. Nobody outlawed motorboats or jet skis. Retirees and snowbirds weren't banned from traveling to the Florida coast. Beaches remained open.

So, what saved the Florida manatee? A combination of things: awareness campaigns, as well as collaboration between government, businesses, and nonprofit organizations. While it's true that manatees were protected at the federal level under the Endangered Species Act, the State of Florida was largely left to enforce that protection. Postage stamps and special-edition license plates spread the word about the manatee. Local governments posted warning signs and set speed limits for boats traveling through waters known to be popular manatee hangouts.

And it worked.

Figure 4.1. *A warning sign for boaters in Miami, Florida.*
Source: https://www.goodfreephotos.com/united-states/florida/miami
/florida-miami-manatees-here.jpg.php.

It may sound strange, but in our efforts to reduce emissions and fight climate change, we'd do well to take a page from the "save the manatee" playbook. Instead of examining all our possible options, we've tended to latch on to one or two single, obvious solutions, like eliminating ICEs, and going all in on EV technology.

It's admittedly hard to argue against measures that could "save our planet" and preserve plants, animals, and resources for generations to come. It's safe to say that, regardless of most political beliefs, religion, nationality, and other typically divisive factors, people around the world are justifiably upset at the rapid environmental destruction humans have caused over the last century. Species are dying at a rate of at least 1,000 times the natural extinction rate,[3] deforestation has left us with only 38 percent of our original forests,[4] and air pollution has become such a threat in certain parts of the world that people don masks as part of their daily routine.[5]

In the discussion regarding electric car adoption, people tend to assume EVs are a highly efficient way to reduce carbon emissions—a forerunner in "must-have" anti-global-warming technologies.

But are they? And if we could reduce the same amount of carbon using a cheaper alternative, wouldn't it be our responsibility to choose *that* option over EVs?

If the goal is to reduce total emissions at all costs, we need to make sure we have all the facts instead of simply grabbing the low-hanging, battery-powered fruit.

Putting a Price on Carbon

Nobody wants to see future generations condemned to short lives on a toxic, hostile planet.

Who doesn't want to believe that there is an easy, accessible solution to such a complex problem? Thus the world has latched on to EVs as a "magic bullet" for mitigating emissions. Consider the words of Norwegian politician Lars Andreas Lunde in an October 2015 interview with *The New York Times*. To be successful in reducing greenhouse emissions, Lunde said, "a large share of the new cars have to be electric . . . it has to be more expensive to pollute than to use environmentally friendly fuels."[6]

Lunde's statement is remarkable only in that it is completely unremarkable. This quote could have appeared in any newspaper. It could have come from almost any politician from practically any country. Around the globe, the prevailing view is that EVs are the best possible path toward an emissions-free future.

Unfortunately, current research doesn't seem to back up this sentiment. Despite vastly different methodologies, author viewpoints, statistical analyses, and a priori assumptions, studies on the cost-effectiveness of EVs in reducing emissions show a surprising—and surprisingly cohesive—trend: Almost universally, EVs are a *very* expensive way to reduce greenhouse gas emissions.

For example, a 2018 report published by Exxon compared the cost of CO_2 abatement across several technologies. The report rated technologies by the cost to eliminate one ton of CO_2 from the atmosphere (see figure 4.2). Wind came in at about $100 per ton, solar at about twice that. Electric cars came in at a shocking $700 or more to eliminate just one ton of CO_2 from the atmosphere. (Of note, in this study, the cost for CO_2

abatement for improved gasoline vehicles was actually *negative,* because of the realized gas savings.)[7]

Emissions - projections
Average U.S. CO_2 abatement costs clarify best options

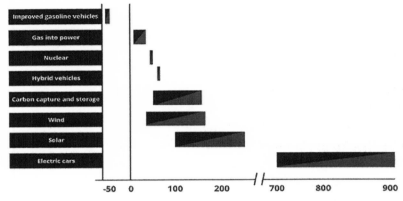

Range of 2016 costs to eliminate one tonne of CO_2 (dollars per tonne)

Figure 4.2. *The cost of CO_2 abatement.*
Source: ExxonMobil, http://cdn.exxonmobil.com/~/media/global/files
/outlook-for-energy/2018/2018-outlook-for-energy.pdf.

For anyone who suspects that this research might be, well, less than neutral given the source, Exxon is not alone in its findings: Earlier studies by Research Gate and the European Commission's Directorate-General Climate Action both agree. Research Gate suggests greenhouse gas abatement costs using plug-in hybrid cars are currently at more than $490 per ton; pure battery-powered cars fare much worse, currently at more than $2,300 per ton of CO_2 abated. A project funded by the commission noted it was incredibly difficult to calculate abatement costs for EVs, but hazarded a guess at $1,237 per ton, at least until 2020.[8]

What's more, costs can vary significantly from region to region. A McKinsey & Co. study on greenhouse gases in Switzerland lists EV abatement cost at somewhere over $124 per ton.[9] In 2015, a report prepared for the Philadelphia Mayor's Office of Sustainability showed plug-in hybrids cost $90 per ton in Philly,[10] and a joint study between Stanford

University and the Precourt Institute for Energy Efficiency found it's approximately $89 per ton in California.[11]

Those numbers certainly sound better than $400-plus—but is $90 per ton actually "cheap"? How do we know?

Greenhouse Gases and Tough Science

Digging into the science of which technologies yield the most cost-efficient emission reduction is not for the faint of heart: Studies are dense, highly technical, and fraught with unavoidable assumptions. When projecting the cost-effectiveness of carbon reduction options, researchers are faced with scores of choices, from how to calculate the true cost of a new technology to how to calculate the true emissions over the lifetime of an EV.

Moreover, it's not even as simple as determining which technologies reduce the most CO_2 per dollar. You must also account for other greenhouse gases (GHGs), such as methane and nitrous oxide, which are significantly nastier for the environment than carbon dioxide. When it comes to global warming, a single methane molecule does as much damage as 34 carbon dioxides, and nitrous oxide knocks methane out of the park—coming in at 296 times the effect of a single CO_2 molecule.[12]

Bloomberg has predicted the cost of CO_2 abatement in the year 2030. Taking the next decade to overcome some initial technology barriers, their marginal abatement cost (MAC) curve predicts the cost of pure EVs landing between $50 and $100 per ton, plug-in hybrids scoring between $0 and $50, and full hybrids falling into the negative.[13] (Full hybrids don't actually plug in, so neither a charging infrastructure nor ultra-high cost batteries are required.)

So, wait, that doesn't sound too bad, right? Does that mean we have the green light? Remember, it depends on your target emissions goal. The higher the emissions reductions goal, the more expensive technologies you should theoretically be willing to pursue.

Stanford suggests that we should put on hold anything with a cost above $70 per ton, and immediately implement only those with a cost less than $35 per ton.[14] RepuTex suggests a cost of $20 to $60 per ton to be reasonable.[15] Berkeley uses $50 per ton as a benchmark in its study.[16] One Canadian study argues that $25 per ton is the magic number, as that is the cost of global warming—anything higher, and it would be cheaper to let the world temperature rise.[17]

Many studies have calculated the economic impact of a warming atmosphere. Analysts use computer models to project the changes in the climate around the world, and then they use economic models to assign a value to those changes. A survey of over 200 estimates of the economic consequences of climate change reported that the median of peer-reviewed studies using a 3 percent discount rate is $20 per ton of CO_2, and with risk adjustment this rises to $25 per ton.[18]

It's also possible to use the price of carbon as a benchmark to indicate how much we should be willing to spend on emissions measures. The price of carbon is a carbon tax that has been levied in dozens of countries and regions; it is the dollar amount that must be paid for the right to emit one ton of CO_2 into the atmosphere. The argument goes like this: If the measure to reduce carbon is going to cost more than the carbon tax, it would be simpler to raise the carbon tax than to introduce the measure. This carbon tax is already levied successfully in many situations.

For example, the coal-to-gas mix in the United Kingdom shifted by 20 percent in 2016 after regulators doubled the minimum carbon price to about $25 per ton.[19] In just four years, the country leaped 13 places in a global low-carbon electricity league, rising from a rank of twentieth to seventh, the fastest ascent of any country.[20]

So if you used the carbon tax as your measuring stick, you wouldn't want to pay more than $25 per ton for abatement measures in the United Kingdom.

Beyond EVs: Alternate Routes to Emissions Abatement

Luckily, there are other ways to reduce carbon in the transportation sector, including many low-cost and even negative cost options for GHG emissions reduction, without having to develop EV technology.

The Transportation Energy Futures study, a Department of Energy project on underutilized GHG reduction strategies, proposes a strategy to reduce GHG emission and petroleum use in the transportation sector by 80 percent. One option not on the list? A widespread transition to EVs. Instead, the project relies on making small changes to the ways we use ICEs. Their strategies include relatively inexpensive, easily implemented things like reducing the speed limit, improving vehicle efficiency, and using mass transportation, car-sharing services, and even carpooling.

In other words, it may very well be that the key to curbing and reducing emissions lies in several small, localized measures: changes in behavior and habits, awareness campaigns, and infrastructure adjustments—strategies that worked out quite well for the once-endangered manatee.

A report published by the International Association of Traffic and Safety Sciences provides an exhaustive list of options, including park-and-rides and telecommuting. The report also provides options to reduce traffic, such as traffic signal control, electronic toll collection, commercial vehicle weigh-in-motion, and ramp metering (to help efficient merging on highways).[21]

While traffic congestion may not be the first thing that comes to mind when considering GHGs, it is actually a significant contributor. For example, traffic in the city of Los Angeles generates an estimated 3.4 megatons of CO_2 per year; this figure represents approximately 3.5 percent of the 170 megatons of CO_2 emitted by California motor vehicles each year.[22]

It's unrealistic to expect that traffic signal control and ramp metering will eliminate all traffic congestion in California, but if system improvements could cut them in half, it would be possible to achieve a GHG emissions reduction of approximately 1.75 percent of the total 170 megatons emitted. It's important to keep in mind, though, that this is not a one-size-fits-all solution: While easing congestion would likely make a huge difference in, say, Los Angeles, Houston, Seattle, or any large, commuter-heavy city, smaller cities and towns with less severe congestion wouldn't see much benefit.

But there are other measures that can make a difference, no matter where you live. Even something as simple as keeping your tires properly inflated can reduce GHG emissions and provide fuel economy benefits of 1 to 3 percent.[23] Unfortunately, according to a random-sample survey by the U.S. Department of Transportation, most tires on American roads are underinflated, and only one in four drivers know how to determine their vehicle's proper tire pressure. The takeaway? It's possible that, at this time, a national tire pressure awareness campaign could get us more GHG-reduction bang for our buck than a state- or federally funded push for EV technology.

Of course, that doesn't mean that this will always be the case. And it doesn't mean that we should abandon EVs simply because they don't provide a quick fix. By doing this, some analysts caution, we may be missing the bigger picture.

In a recent interview, Jack Barkenbus, a senior researcher at the Climate Change Research Network, said, "I have no doubt that the development of EVs is not now the most cost-effective method for removing CO_2 emissions. But the issue raised is flawed on many levels. First, EVs have other benefits leading to their development than just CO_2 removal (e.g., local air improvement, reducing oil imports, etc.). Also, the effectiveness of EVs as a climate change measure depends on the energy mix of electricity producers, and, this varies across the globe and within countries. The important point though is that in the long run electricity producers are reducing the CO_2 levels of their product and hence EVs, as a climate control measure, should become more cost-effective in the future."

So, do EVs have a cost-efficient chance of benefiting us after all?

The answer is . . . it depends.

In a December 2016 presentation on the future of EVs and oil demand, BP economists Spencer Dale and Thomas D. Smith pointed out that it's nearly impossible to pin down the exact emission savings associated with EVs: "Amongst many other things, it depends on: the distribution of EVs across different parts of the world and the structure of their power systems; the pace and extent to which the fuel mix used within different power sectors changes and/or Carbon Capture and Storage (CCS) is deployed; and even, potentially, on the time of day at which EVs are typically recharged (since in some countries the relative importance of renewable energy in power generation varies significantly through the day)."[24]

And there it is. The missing link.

Location, Location, Location

For the time being, the old real estate adage holds true for EV technology and emissions reduction: The benefits of EVs largely depend on the city, state, and even country in question. National energy infrastructure makes a difference. A country's predominant power source (i.e., coal, solar, natural gas, etc.) also matters; a "cleaner" power mix (i.e., less power generated from coal and ore from cleaner energy sources) would lower the abatement cost as the CO_2 abatement levels of EVs would increase. Other factors include local gas prices (the higher the gas price the more cost-effective EVs become in terms of carbon reduction), and local issues with pollution and smog also come into play.

One thing is clear, though: EVs are more effective when local grids are cleaner, and have other specifically local benefits, which suggests that regional, state, and city adoption strategies make more sense than federal EV subsidizing.

In California, for example, emissions for EVs are comparatively low (as low as 100 grams per mile of greenhouse gases) because the grid has a large percentage of renewable electricity. But in the Midwest and South, where coal fuels the bulk of electricity generation, a hybrid gas-electric car actually produces *less CO_2 than a pure electric car.*

In Minnesota, which is dependent principally on coal and other fossil fuels for power, an electric car would actually emit 300 grams per mile of greenhouse gases. "As a result," according to David Biello writing for *Scientific American*, "some researchers suggest that a regional approach to clean vehicle standards makes more sense than national standards that effectively require electric cars across the board. Minnesota could go for hybrids and California could go for electric vehicles."[25]

The Department of Energy (DOE) actually publishes comparative charts per state showing both the split of energy resources used to generate electricity (e.g., wind, solar, coal, natural gas, nuclear, hydro, geothermal, and other sources) along with respective CO_2 emissions for regular gasoline, diesel, hybrid, and pure EVs. The interactive charts are revealing. They show that on a national scale, EVs, hybrids, and plug-in hybrid-gas vehicles emit around 50 percent less carbon dioxide annually than gasoline engines do (you can view the charts at www.afdc.energy.gov/vehicles/electric_emissions.php). The comparative bar charts allow any viewer considering an EV, hybrid, or conventional car to make an educated choice.

As local regions choose the GHG emission reduction technologies that work best in their areas—through careful planning based on actual scientific research, not political feel-good measures—they will begin to share the knowledge of best EV adoption strategies. Assuming EVs are a good local solution, they will become more and more widely adopted, naturally, as cities learn from their neighbors and charging infrastructure grows organically.

In fact, if the immediate goal is to reduce global carbon emissions in the short term, subsidizing EV sales with federal funds is questionable, at best. There are clearly far cheaper ways to reduce carbon. And there are likely more efficient ways: Although EVs are often hyped as game-changers in the battle against emissions, the data tells a different story.

In BP's 2016 Energy Outlook report, Spencer Dale and Thomas Smith provided a detailed look at the realities of EVs and emissions reduction. "Under a scenario with no EVs, the world's cars are estimated to produce 3100 million tonnes of CO_2 in emissions per year by 2035. That figure falls by 75 million tonnes based on the amount of EV switching assumed in BP's Energy Outlook, and by a further 405 million tonnes under a scenario with an additional 380 million electric vehicles," they wrote. "To put those figures in context, total CO_2 emissions from global energy consumption in 2015 amounted to 33.5 billion tons."

Although their report notes that a total reduction of 480 million tons per year represents "significant savings," Dale and Smith were quick to point out that widespread EV adoption is just one way to achieve such results: "A 4 [percent] shift today in the global power sector away from coal-fired generation to gas generation would lead to broadly the same saving in CO_2 emissions."[26]

For now, any regulation that pushes heavy EV subsidies in the name of climate improvement should be heavily scrutinized. The benefits of EVs may lie in climate improvement someday, but today other technologies win out when measured by that barometer. Individual citizens, along with state and local governments, should be the ones who decide whether EV benefits are right for them; federal governments should not try to enforce a one-size-fits-all solution.

Perhaps one day in the not-too-distant future, another bombshell announcement will make the headlines: GHG emissions have been slashed; the climate change debate is over.

But it won't happen overnight, to be sure. And it won't happen without widespread buy-in and a willingness to work together, where we examine all our options and make the best choice.

Perhaps that will include EVs. Perhaps not.

In the end, though, it's not how we get there that matters. And that's something on which we should all be willing to agree.

CHAPTER 5

THE KING IS DEAD.

LONG LIVE THE KING

A discussion on EVs and emissions could not be complete without exploring how much oil electric cars will displace. After all, every EV sold reduces our need for oil. So how much downward pressure is oil demand likely to see?

Some claim, quite a lot.

Of course, the death of the oil industry has been a favorite topic since . . . well, since we started to really use oil. As early as 1909, newspapers were predicting the end of oil was in sight. On July 19 of that year, Pennsylvania's *Titusville Herald* ran an article citing U.S. Geological Survey predictions that all natural gas would be exhausted by 1934, petroleum and iron by 1939, and coal by the middle of the 21st century. In 1937, Captain H. A. Stuart, director of the naval petroleum reserves, was sure that oil couldn't possibly last more than 15 years. "That is a conservative estimate," he told the Senate Naval Affairs Committee. "We have been making estimates for the last 15 years. We always underestimate."[1]

In the 1950s, we were introduced to one of the most famous forecasters of oil's demise: Marion King Hubbert, a geoscientist at Shell research
lab, who proclaimed that oil reserves would peak in the 1970s and be
depleted just a few decades later. (The peak point of maximum production came to be known as the peak oil theory.) To come to this conclusion,
Hubbert calculated the rate of consumption versus available resources
and showed clear as day for anyone to see, that oil would clearly run out.

After all, it's common sense. If you have a limited supply coupled
with growing use of a product, at some point the supply will hit rock
bottom (literally).

When the 1970s brought us petroleum and gas shortages, embargoes,
and a quadrupling of petroleum prices, Hubbert was called the "oracle"
of oil.[2] Some oil companies began to panic, with the CEO of Mobil saying
in 1977, "The oil business has come to maturity, and with this maturity
comes a new set of challenges . . . oil companies have no other choice.
They must diversify or go the way of the buggy-whip makers."[3]

By the 1980s, everyone could see that "Hubbert's pimple" was following Hubbert's bell-curve prediction almost exactly (see figure 5.1).
U.S. oil production had peaked in 1970, then steadily declined . . . and
declined . . . and declined.

The 1980s were a very bad time to be in oil.

In retrospect, the downward slide looks more like a blip on a big-
ger-picture exponential oil growth curve. Today, we use more oil than
ever, with approximately 45 million barrels a day in transport and 96
million barrels per day in global use for all applications.[4]

Still, the futurists aren't done pronouncing the death of oil.

More recently, biofuels, hydrogen cars, and Segways (if you can
believe it) have all been used to create endgame predictions for the oil
and gas industry.[5] Even the Internet was used to predict a decline in gas
sales in the 1990s—people wouldn't need to travel if they could telecon-
ference; they wouldn't need to ship packages if they could get documents
via email.

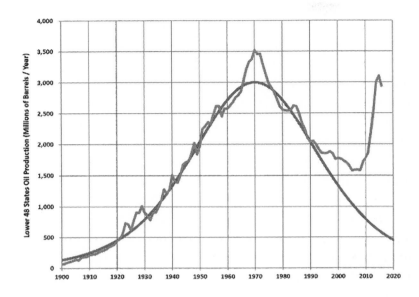

Figure 5.1. *Hubbert's Pimple. Today it looks like a mere blip on a larger growth curve.*
Source: https://upload.wikimedia.org/wikipedia/commons/f/f2
/Hubbert_Upper-Bound_Peak_1956.png.

In recent years, the newest harbinger of death for oil is—yep, you guessed it—the electric vehicle. Just look at the recent headlines:

- "Electric Vehicle Sales Foretell a Big Oil Crash"— EcoWatch, July 2018

- "Oil Prices at Risk as More Electric Vehicles Hit the Road"—Fox Business, October 2018

- "Big Oil Makes a New Attempt to Kill Electric Cars"—InsideEVs, December 2018

- "Big Oil Taps into Electric Era"—*Automotive News*, March 2019

- An October 2017 article on hybridcars.com announced, "Oil Demand Will Plummet by 2025 Due to Electric Cars." The first sentence proclaimed, "Thanks to electric vehicle adoption, global oil demand could be reduced to 3.5 million barrels a day by 2025." They cited Barclays as the source, then continued with, "The 3.5 million barrel figure nearly matches Iran's daily output . . . a doomsday threat to OPEC."[6]

But how much is 3.5 million barrels a day? It sounds like a lot. But when you realize that our 2016 total petroleum usage was 96 million barrels a day,[7] and, according to Statista, the 2019 total was 101.1 million barrels a day, the earlier claims start to sound a lot like hyperbole.

Old Habits Die Hard

Why does the demand for fossil fuels continue to rise despite attempts to replace it with cleaner sources? Notwithstanding what some may think, it is not the evil plot of senior oil and gas executives and crooked politicians, living in penthouses, smoking cigars, and plotting the destruction of the earth in country club steam rooms and backroom secret societies.

It's because oil is really, really good at what it does.

Although many might think it would be great to just replace all fossil fuel energy with renewables, the reality is that, assuming we all want to continue using energy, the feat of going to an all-renewable energy grid is impossible.

Not improbable. Not hard. *Impossible.*

At least, it is impossible today. That is not to say that some genius engineer won't ever invent cold fusion or some other equally life-changing scenario plays out. Now, however, there is simply no clear path that leads to all-renewable energy sources.

To help explain why, the scientists at SRI International developed a unit that allows us to compare energy sources apples to apples. It's called a CMO and stands for cubic mile of oil equivalent (i.e., the amount of energy needed to replace one cubic mile of oil). Today we use about three CMOs per year.

To replace just one of those CMOs per year with renewables, you would have to install—

- 4 Three Gorges Dams each year for 50 years, *or*

- 32,850 wind turbines each year for 50 years, *or*

- 91,250,000 rooftop solar photovoltaic panels, each year for 50 years.[8]

If that doesn't provide enough pause, consider this:

- Each of the Three Gorges Dam costs roughly $30 billion, displaces 1.25 million people, and floods 632 square kilometers.
- Each wind turbine costs $2 million and needs about .16 square kilometers, not to mention that you also must have wind.
- Each solar panel must be installed in a sunny location and costs about $15,000.[9]

Even if we could stomach the costs associated with this massive infrastructure overhaul, there are other setbacks as well. We could face very real barriers in the availability of materials needed and the cost of building out new transmission lines, which can run into the billions even for short regional lines.[10]

Finally, the amount of land required for the implementation of such a large undertaking would be stomach churning. Using one plan proposed by Mark Jacobson and others, the United States would require 46,480 solar photovoltaic plants to meet the 100 percent Clean and Renewable Wind, Water and Sunlight vision. By one estimate that would take up to 650,720 square miles, or almost 20 percent of the lower 48 states.[11] Would not a significant portion of this land have come from currently undeveloped lands—lands we are saving for wildlife; national, state, and local parks; and ecotourism?

There must be a better way.

It's hard to dispute. We use a LOT of energy—about 241 million barrels of oil equivalent per day, of which about 210 million barrels of oil equivalent come from hydrocarbons.[12] That is a shocking quantity of hydrocarbons we need to replace with cleaner sources.

Germany has been called "The World's First Major Renewable Energy Economy."[13] They are pushing hard on renewables and making some impressive strides, including three giant wind plants in the North Sea—together rivaling the power of a nuclear reactor. Germans achieve 38 percent of their power from renewable sources.[14] And yet, the nation still generates 55 percent of its electricity from fossil fuels.[15] Perhaps even more shocking, Germany has not succeeded in reducing its carbon emissions nearly as much as you would expect. In 2014, the country vowed to cut these emissions by 40 percent by 2020 . . . only to admit in 2018 that 32 percent might be a more realistic goal.[16] Although Germany did see a precipitous decline from 1990 to 2009 (from 1,251 to 907 CO_2 equivalents in millions of tons), progress seems to have plateaued—it may even be going backward. In 2016, it had virtually the same level of CO_2 emissions as 2009, and actually *higher* emissions than the previous two years.[17]

What seems to be the problem? Because solar and wind are intermittent sources of power, Germany keeps coal plants running—even when the grid doesn't need the power.[18] In fact, in 2016, seven out of ten of Europe's biggest polluters were German lignite (brown coal) stations.[19]

Perhaps surprisingly, from 2005 to 2015, the United States, which pushed for greater natural gas usage along with wind and solar, saw a larger decrease in power generation's CO_2 intensity than Germany—a 20 percent decrease in the United States as compared to 10 percent in Germany.[20]

The reality is, it's just not that easy to replace fossil fuels—coal, natural gas, or, of course, oil.

Driving on Empty?

In his book *Power Hungry*, Robert Bryce writes, "If oil didn't exist, we would have to invent it. It is the most flexible substance ever found."[21]

Oil is particularly useful in the transportation sector, where it supplies 95 percent[22] of the energy for road transport. (That's roughly the same percentage as 1971 despite the more than $250 billion[23] spent since then in alternate energy solutions.)

Just as we have had trouble replacing fossil fuels on the electric grid, so too do we see hydrocarbons' stubborn stronghold in transportation. Alternate resources have found it hard to compete with oil's energy density, ease of transport, and global ubiquity. And EVs have a *long* way to go before they make up a substantial portion of the global fleet, even if they become cost-effective, or popular.

Let's consider the hurdles.

We know that, today, there are about 2 million electric cars tooling down the road around the globe.[24] We also know that some 79 million cars were expected to be sold worldwide in 2019 alone.[25]

From today to 2040, the global fleet is predicted to go from 1.1 billion vehicles to 2 billion.[26] That's almost doubling the number of cars on the road. (Most of these new vehicles will be in Asia, which is already home to one out of every three of the cars on the planet. By 2040, that number will be closer to one out of every two.)[27]

Finally, on average, an ICE vehicle's lifespan is approximately ten years. So for every non-EV sold today, it will be about ten years before that car leaves circulation.

It doesn't take a mathematician to realize this: To surpass ICEs in numbers, EVs must do more than compete neck and neck—they have to annihilate the competition. Anything less, and they'll be playing catch-up for a very long time.

The task of replacing 45 million barrels of oil per day in the transportation sector is momentous. (EVs use mostly natural gas and coal as electric power sources, so they are certainly not emissions-free, especially in China, which still generates 69 percent of electricity from coal.)[28]

The task of replacing all the hydrocarbons required to serve a growing *worldwide energy demand across all sectors* is even more momentous.

Energy Wins, Energy Woes

Worldwide demand for energy continues to grow, especially with urbanization in China and India. According to the 2019 Energy Outlook, world population is expected to grow from 7.3 billion to 9.2 billion by 2040, driving energy demand up with it. Way up: Thanks to increasing prosperity in developing nations, we could see demand up by a third.[29]

As global living standards improve, the 1.2 billion people around the world without any access to electricity—*at all*—will begin to come online.

And that's a good thing.

Currently those without electricity are among the poorest on the planet, living without easy access to clean water, food, or medical services. It may be in vogue to romanticize "the good ol' days" when our forefathers did just fine without lights and wall plugs. But it's a tragedy when you consider that in India, which houses an estimated one-third of the world's population living without electricity, approximately 1.83 million children under five years of age die each year. More than two-thirds of those children die within the first month, with 90 percent of those deaths being caused by something preventable such as pneumonia and diarrhea.

There is a reason electricity is so popular. Electricity makes our lives safer, not to mention more productive, and much more comfortable.

As the poorest of the poor begin to move into lighted homes and gain access to hospitals with dependable electricity, the currently impoverished will begin to move into the middle class. By 2030, the global middle class will more than double, reaching 5 billion people. (Exxon's outlook put this currently closer to 2.3 billion.) That means more air-conditioning, more computers, more cell phones, and more appliances.

All of this means that we need . . . you guessed it . . . more energy.

To meet the escalating demand—while continuing to reduce emissions—we are going to have to pull out every trick we know.

Luckily, our scientists have been hard at work, and their work is paying off. Our energy intensity (the amount of energy used per unit of economic output) has reduced about 1 percent per year on average since 1970 and is expected to reduce about 2 percent per year from 2015 to 2040.

That means we can do more with less.

We will meet our increased energy need, according to the Exxon outlook, with about 700 quadrillion BTUs versus the 1,200 quadrillion BTUs we would need without these energy savings.

More good news: Between now and 2040, renewable and nuclear power are projected to be the world's fastest-growing forms of energy, and coal moves down. Still, petroleum remains the largest source of energy, even as its share of energy is predicted to decline from 33 percent (2015) to 31 percent (2040).[30]

When Hubbert published his peak oil theory, the concern was that the world would run out of fossil fuels. It's interesting to see how the conversation has shifted.

Today's editorial rarely talks about peak supply. Instead, the word of the day is *peak demand*. When will the hydro-powered tide turn? When will Germany's great North Sea windmills finally usher in the winds of change? When will the sun shine on our solar-powered ambitions?

In a recent report, Carbon Tracker compared oil demand predictions from IEA, Shell, BP, ExxonMobil, and OPEC. All except the IEA 450 scenarios forecast *increasing* oil demand at least through 2035.[31]

In late 2017, Shell published their Oceans and Mountains scenarios for oil demand. In their Oceans scenario, liquid fuels still deliver 70 percent of the fuel to passenger vehicles, and oil demand grows until 2040, when it reaches a long plateau. In the Mountains scenario, liquid fuels for passenger transport decline after 2035, and oil demand peaks in midcentury. By 2060, oil is a smaller portion of the energy mix, but it's still larger than solar, biomass, or wind. Coal, too, remains a big player.[32]

Since no one is likely to turn off the spout in 2060, it means that oil will be around for a while.

On the Road Again

Clearly oil is affected by myriad factors: a push toward lower CO_2 emissions; future energy demand; fuel efficiency; China and other emerging market economies; fuel switching to natural gas and solar; and oil-to-gas switching.

Reduction in oil subsidies could also displace oil in the future. For example, according to the IEA, "At the extremely low gasoline prices prevailing in Saudi Arabia today, an investment in a more efficient car (consuming half the gasoline of an average Saudi Arabian's car) would pay back only after almost 20 years."[33]

Electric cars are likely to be a tiny portion of what affects oil and gas demand.

But, just for the sake of argument, how much oil *could* EVs displace?

The supermajors, at least, see a world where fossil fuels play less of a role in transportation. French oil company Total Global is now saying EVs may constitute almost a third of new-car sales by the end of the next decade.[34] ExxonMobil boosted its 2040 estimate to about 100 million from 65 million. BP anticipates 100 million EVs on the road by 2035, a 40 percent increase in its outlook compared with a year ago. Statoil ASA, the Norwegian state oil company, says EVs will account for 30 percent of new sales by 2030. Even OPEC has raised its 2040 EV fleet prediction—to 266 million from the 46 million it anticipated a year ago.[35]

With that many new EVs, surely the oil demand will go down? Surprisingly, by most accounts, it doesn't. Or at least not anytime soon.

According to an EIA forecast, the demand for petroleum-based fuels for passenger vehicles continues to *rise*, despite the growing population of EVs on the road. Remember that doubling of cars on the road problem we talked about earlier?

Interestingly, most studies indicate that efficiency increases in petroleum cars will cause more of a reduction in oil demand than will the introduction of EVs. The 100 million increase in electric cars by 2035 will reduce oil demand growth by just 1.2 million barrels per day. By comparison, this is around a tenth of the impact of the gains in vehicle efficiency.[36] The average fuel economy of new cars will rise from 30 mpg to 50 mpg.[37]

Wouldn't it be somewhat ironic if it were the traditional automakers who turned out to be the real CO_2 savers after all?

Navigant Research estimates EVs displaced 2.1 million barrels of oil in the United States from 2011 to 2014. Yet to accomplish this same savings, ICEs need only improve fuel efficiency roughly .08 percent in four years—that's nothing when you consider that the U.S. CAFE standards will increase vehicle fuel efficiency 22 percent in the next ten years.[38]

Back to the Future: 2060

How will we remember today in 40 years? We lay out our predictions, and we work hard to see them through.

How are we doing? Perhaps not as poorly as it seems. Although the challenges of reducing CO_2 emissions are much more complicated than some people realize, we innovate daily.

We have made, and we will continue to make, great strides.

If one thing *is* certain, it is that 2060 will surprise us. Don't believe it? Just think about 1980. How many revolutions have we seen since then?

In 1980, computers were relatively new and only available to people with large bank accounts.

Seat belts were rarely used, and car seats (along with car seat laws) were practically unheard of.

If you wanted to buy something, you had to physically go to the store.

Most homes only had one TV, and that TV had a whopping three to four channels.

Telephones were attached to the wall. There was no such thing as call-waiting, redial, or multiple lines.

Research projects had to be done in libraries, using card catalogs, microfiche machines, and actual books.

Will the next revolution be the EV revolution? Will fossil fuels finally have their last hurrah? The fact is, we don't know. However, we do know that the challenges are real, and real technology will need to be behind any large-scale changes.

As far as EVs single-handedly pushing out oil, well—for now, at least, that part of the story is more science fiction than science.

CHAPTER 6

ARE SUBSIDIES "LUDICROUS"?

The jury is in. The verdict is delivered. EVs are currently a very expensive way to reduce carbon, *and* EVs are unlikely to put a dent in oil demand. However, subsidizing EVs isn't just a questionable environmental effort or a poor investment of taxpayers' dollars; it turns out that subsidies are also bad for business.

One of subsidies' most surprising critics is none other than Elon Musk, the South African–born billionaire, founder of the legendary SpaceX and of leading electric car manufacturer, Tesla. During Tesla's 2017 first quarter Q&A call, Musk bemoaned, "Over the years there's been all these sort of irritating articles like Tesla survives because of government subsidies and tax credits . . . It drives me crazy . . . all those things would be material if we were the only car company in existence.

We are not. There are many car companies. What matters is whether we have a relative advantage in the market."[1]

Tesla is in an awkward position when it comes to government subsidies: Early on in its relatively short life, the company received $465 million from the government's Advanced Technology Vehicles Manufacturing (ATVM) Loan Program.[2] The funds helped Tesla develop the Model S, a luxury vehicle starting at $68,000 in 2017 and boasting features like a 17-inch touchscreen display, retractable door handles, and available high-end extras like "Sonic Carbon Slipstream" wheels and the appropriately named "Ludicrous Speed" option, which enables the vehicle to accelerate from 0 to 60 mph in 3.2 seconds.

Whatever position you may hold on subsidies in general—whether you consider yourself liberal or conservative; whether you vote Republican, Democratic, Green, Libertarian, or Independent—it's hard to make the case that U.S. taxpayers should foot the bill for the production of luxury cars marketed to the wealthiest part of the population.

You can only imagine the outcry if, say, the government was subsidizing the production of BMWs or Cadillacs. Or, you can take a look at the very non-imaginary outcry in Alabama in the late 1990s, when state lawmakers in Alabama subsidized the construction of a Mercedes-Benz plant in Vernon,[3] where the population is just north of 2,000 and the median income is around $34,000.[4]

And yet, many people argue that because Tesla produces EVs—even though the EVs in question, starting now at $80,000, come with a sticker price that's well over twice the median income in Vernon, Alabama—the company should get a pass. Some EV proponents point to subsidy programs in other countries as proof that, when it comes to promoting EVs and reducing our dependence on foreign oil, subsidies aren't so bad, after all.

EV Subsidies: Today's Regulatory Framework

The loan that helped Tesla develop the Model S is part of the Energy Independence and Security Act (EISA) of 2007, which was signed into law by then president George W. Bush on December 19, 2007:

An Act

Dec. 19, 2007
[H.R. 6]

To move the United States toward greater energy independence and security, to increase the production of clean renewable fuels, to protect consumers, to increase the efficiency of products, buildings, and vehicles, to promote research on and deploy greenhouse gas capture and storage options, and to improve the energy performance of the Federal Government, and for other purposes.

Figure 6.1. *Energy Independence and Security Act (EISA) of 2007.*

EISA, which for the most part received bipartisan support, had admirable goals. Authorized under section 136 of EISA, the ATVM Loan Program was meant to promote research that would help the United States achieve greater energy independence and make vehicles more efficient.[5] In June 2009, the Obama administration announced it would award $8.5 billion in loans to help companies like Tesla research and innovate—without budget constraints.

The recipients of the first three loans under the program were Tesla, Nissan, and Ford. Energy Secretary Steven Chu said the program would help the United States meet many of the goals outlined by EISA: "By supporting key technologies and sound business plans, we can jump-start the production of fuel efficient vehicles in America," he said. "These investments will come back to our country many times over—by creating new jobs, reducing our dependence on oil, and reducing our greenhouse gas emissions."[6]

Two years later, in 2011, the Obama administration pledged $2.4 billion in federal grants to support the development of EV technology, with the goal for the United States to become the first country with 1 million EVs on the road—by the year 2015. As of December 2016, we were still only about halfway there.

Other incentives, subsidies, and tax breaks have also been put in place to encourage American consumers to go electric: The passage of the Energy Improvement and Extension Act of 2008 offered up to $7,500 in tax credits for the purchase of a new EV. These tax credits will phase out once each manufacturer sells 200,000 EVs in the United States. Once an EV manufacturer reaches the limits, the tax credit will drop to $3,750 for six months, and then to $1,875 for another six months until the credit runs out completely.

It's true that, to some degree, government funding has helped accelerate the development of EV technology. It's also true that Tesla is not the only company that received conditional loans under the ATVM Loan Program: Ford, Nissan, and the now defunct Fisker Automotive also received funds.

According to the Department of Energy, both Tesla and Nissan have repaid their loans in full. Nissan repaid its loan in September 2017 and Ford is scheduled to repay in September 2022. Tesla repaid—with interest, and nine years early—in May 2013.[7] And despite his company receiving government subsidies, Elon Musk has been an outspoken critic of government funding. Yes, funding helped in the short term. But long term, he says, government support is ultimately hurting him and stifling competition.

As Musk stated in his quarterly financial results conference call, "In fact the incentives give us a relative disadvantage. Tesla has succeeded in spite of the incentives not because of them."[8]

To understand why Musk is so frustrated with subsidies, it is important to understand the zero emissions vehicle (ZEV) program. ZEV, which is active in California and nine other states, mandates auto manufacturers produce a certain number of electric cars, or pay a $5,000 fine for each EV they do not produce. An auto manufacturer can avoid this fine by purchasing ZEV "credits" from manufacturers (like Tesla) who earn extra credits for producing more EVs than their quota.

These credits are not sold at $5,000. Instead, due to Tesla's oversupply, they are sold by Tesla to competitors at a discount around $1,600.[9] On one hand, this provides Tesla with an essentially free stream of income. On the other hand, Tesla is selling something for $1,600 that is worth $5,000 to its competitor.

"If General Motors or Ford or somebody else makes electric vehicles, they get to monetize their ZEV credit at 100 cents on the dollar," explained

Musk.[10] However, because Tesla is generating more credits than it needs, credits become worth more to their competitors than to Tesla.

Meanwhile, a traditional automaker who may believe EVs are commercially unviable is forced to make them anyway. Under current government regulations, automakers are manipulated into producing more EVs than there is demand for. For example, ZEV mandates automakers must sell at least 15.4 percent EV (or hydrogen) cars by 2025.[11] If the demand is not there by 2025, these cars will have to be sold at a loss.

If that's not convoluted enough, the way regulations are currently structured allows ICE automakers to "game the system," canceling out the very emissions reductions the regulations were intended to create. Consider the Corporate Average Fuel Economy (CAFE) standards, which require that all vehicle manufacturers increase average fuel economy across all models by almost 100 percent and reduce emissions levels by more than 40 percent by 2025. (These gradual increases began to be imposed with 2012 model years.)

The problem lies in the language: "Across all models" means that, in theory, the more EVs a manufacturer produces, the less that manufacturer needs to worry about lowering emissions from its fleet of ICE vehicles. In other words, vehicle manufactures have a pass to churn out gas-guzzling pollution machines—if they produce enough EVs to make up for it. Another problem? "Make" is not the same as "sell": Those clean, green EVs aren't going to do any good if people are still buying their ICE counterparts.

Bad Karma: Fisker's Government-Funded Failure

Elon Musk says Tesla's success is due to the quality of its cars, not government funding—but whether you believe him or not, it's impossible to separate the success from the subsidies. We can't know if Tesla would have skyrocketed to the top without assistance, and, for now, it's too early to tell if the company's current success is sustainable without government funds.

But we can learn something from the spectacular failure of Fisker Automobiles. The Fisker failure is a perfect example of how government-funded innovation can go terribly wrong.

In January 2008, Fisker rolled out the sleek, sporty Karma at the North American Auto Show in Detroit, Michigan. Leonardo DiCaprio was an early fan of the Karma, so were Al Gore and Carlos Santana. Comparisons to Tesla were automatic, and, depending on who you believe, not a coincidence: Just a few months after the Karma made its debut, Tesla filed a lawsuit claiming Fisker had stolen their designs and trade secrets. The suit was settled in Fisker's favor. A year later, Fisker made headlines when it became one of five vehicle manufacturers to receive a $528.7 million loan under the ATVM Loan Program.

Fisker purchased a former GM manufacturing plant in Delaware. Then vice president Joe Biden traveled to the site and waxed poetic about the future: "Imagine when this factory, when the floor we're standing on right now is making 100,000 plug-in hybrid sedans, coupes, and crossovers every single year," he said.[12]

And then, in 2011, news outlets began to tell a different story: Those hybrid sedans, coupes, and crossovers wouldn't be manufactured in Delaware—or anywhere in the United States, for that matter. Although the design work would be completed in the States, the assembly would take place in Finland. This meant that the American people lost approximately 500 assembly jobs—jobs that, to some degree, they had paid for.

Company CEO Henrik Fisker's explanation? We didn't have the manufacturing capabilities.

"There was no contract manufacturer in the United States that could actually produce our vehicle," he told ABC. "We're not in the business of failing; we're in the business of winning. So we make the right decision for our business."[13]

Unfortunately for Fisker, assembling its vehicles in Finland wasn't enough to keep the company in the business of winning—or in business at all, for that matter: In December 2011—just one month after Fisker began delivering its first vehicles to customers—the company issued a recall of all vehicles manufactured between July 1, 2011, and November 3, 2011.[14] The reason: A faulty battery (a problem with the hose clamps made the lithium-ion battery prone to both short circuits and fires). The recall affected 239 vehicles, which comprised nearly all the vehicles shipped to customers, plus most of the vehicles sitting on dealership lots.

In March 2012, Consumer Reports published a scathing review of the Fisker Karma. In addition to smaller issues—from design flaws to engine noise to battery recharge times—the vehicle actually failed during routine testing. "While doing speedometer calibration runs on our test track (a procedure we do for every test car before putting it in service by driving the car at a constant 65 mph between two measured points), the dashboard flashed a message and sounded a 'bing' showing a major fault," the review detailed. "Our technician got the car off the track and put it into Park to go through the owner's manual to interpret the warning. At that point, the transmission went into Neutral and wouldn't engage any gear through its electronic shifter except Park and Neutral."[15]

It went on to remark that the vehicle's failure was something of a milestone: "We buy about 80 cars a year and this is the first time in memory that it is undriveable before it has finished our check-in process."

From there, things just got worse: In October 2012, Fisker halted vehicle production after A123 Systems, the company that manufactured the Karma's (faulty, fire-prone) batteries, went bankrupt. (It's worth mentioning that A123 had also benefitted from government help: In 2009, A123 received a $249 million grant from the Department of Energy as part of its Electric Drive Vehicle Battery and Component Manufacturing Initiative.[16]) A year later, Fisker was purchased at auction by the U.S. unit of the Wanxiang Group.

On September 13, 2013, the Department of Energy posted a lengthy update on its website, outlining Fisker's failures and the status of the loans. "Unfortunately, as has been widely reported, Fisker Automotive has experienced major setbacks in their production schedules and delayed sales that caused them to miss critical milestones laid out in their loan agreement with the Energy Department," the report stated. "After exhausting any realistic possibility for a sale that might have protected our entire investment, the Department announced today that we are auctioning the remainder of Fisker's loan obligation, offering the best possible recovery for the taxpayer."[17]

The update also noted that of the $528 million it had earmarked for Fisker, the U.S. government had only disbursed $192 million by the time the company went under.

But of that $192 million, only a small amount was recouped, leaving U.S. taxpayers $139 million short, with nothing to show for it but an abandoned plant in Delaware and a few hundred fire-prone EVs destined for the junkyard.

While government-funded research can, as Steven Chu said, "jump-start" important discoveries and innovations, it also hinders competition between the technologies and prevents the market from accurately dictating which technologies thrive.

What's more, subsidizing one innovation over another can discourage the development of a potentially superior technology. More promising companies that could have received funding in the private sector may not get the chance if the government is backing their competitors.

This occurs because the backing lowers the perceived risk of government-favored companies, causing them to appear more attractive to investors. Government funding distorts the investment risk to such a degree that even companies that are commercially inviable—such as Fisker—can appear almost irresistible. Case in point, after announcing the support of Energy Department funding, Fisker saw a flood of private investment, raising $600 million before it even sold a car.[18]

Worse, government funding of private companies encourages corruption, and may even make corruption unavoidable. The funding creates a symbiotic—or perhaps parasitic—relationship between the government and its private partners, making it very difficult to avoid providing mutual favors. While companies with strong government connections will receive funding, companies without political connections—innovative companies that may be equally or even more deserving—won't get the support they need.

It's impossible to know what would have happened if Fisker had not received its federal funding. One thing that is clear, though: The U.S. taxpayers wouldn't have footed the bill for the failure.

Perhaps Nicolas Loris, an energy analyst who provided congressional testimony in the Fisker case, best summed up the situation: "Having the federal government provide the loan privatizes the benefits and distributes any potential losses among the taxpayers."[19] If the company is a success, the taxpayers see no financial reward for their investment. If the company is a failure, the taxpayers suffer 100 percent of the loss.

BEYOND EVS: THE SUN SETS ON SOLYNDRA

EVs aren't the only "green" technology championed by the U.S. government—and Fisker isn't the only colossal failure. The Solyndra debacle might be the most (in)famous and controversial example of government subsidies gone horribly awry.

Solar panel manufacturer Solyndra received the first U.S. Department of Energy (DOE) loan guarantee under the stimulus to support its development of unique cylindrical solar panels that would absorb the sun's energy from any direction.[20] Using its loan guarantee along with almost $200 million from private investors, the company opened a new state-of-the-art robotic facility in September 2010 . . . and then abruptly filed for bankruptcy in September 2011. A resulting U.S. Department of Treasury criminal probe investigated whether the company had misrepresented its financial situation when seeking the loan or engaged in accounting fraud.

The Washington Post reported that even when the company's records warned of looming financial disaster, the Obama administration had remained dedicated to supporting the company's cleantech efforts.[21]

The failure of Solyndra was a perfect storm of extremely high-risk, unproven technology, dishonesty on the part of company executives, and poor judgment. In a report issued by DOE Inspector General Gregory H. Friedman, the DOE's due diligence process was "less than effective" and "the Department missed opportunities to detect and resolve indicators that portions of the data provided by Solyndra were unreliable."[22]

continued

Sure, it has the largest price tag on this list, and it signifies a huge federal gamble that lost big-time. In addition to hurting taxpayers, though, DOE loans and loan guarantees also hurt innovators and entrepreneurs: By dictating who should receive capital, they are essentially deciding who deserves an opportunity. A company that the government views as a strong contender will get funding, but companies on the fringes—perhaps companies without celebrity backing or strong connections to Washington—may lose the opportunity.

This serves as a reminder that the government is not well positioned to make decisions about which technologies to invest in. As the World Trade Organization says, "Decisions about what to subsidize often involve technical complexities about which governments lack adequate information."[23]

In other words: Let's keep the innovation in the hands of the innovators.

Norway and EVs: A Misleading Success Story

Norway is often lauded for leading the world in EV usage.

"Norwegians switch to electric vehicles faster than anyone else on the planet," writes Matthew Campbell of *Bloomberg Businessweek*. "More than a third of all new cars are either fully electric or plug-in hybrids, well over 10 times the proportion in the United States. With about 100,000 electrics on the road, Norway (population 5 million) trails only the United States, China, and Japan in absolute numbers. By 2025, the government has suggested, there may be no gasoline- or diesel-powered cars sold in the country."[24]

But here is perhaps one of the biggest examples of misleading EV "success stories." The thing is, the fuel making these impressive statistics possible is the mechanism of government subsidy.

From EV sales tax and vehicle registration fee exemptions to free plug-ins at municipal power stations to a free pass on road tolls and public ferries, government subsidies are the motivating force behind the country's strong EV market. Norway introduced the subsidies in the 1990s in hopes of boosting what was then a very young and less-than-thriving domestic EV industry. When the Nissan Leaf and Tesla Model S became available, larger volumes of people started taking advantage of the government's extensive EV incentives.

The carrot of subsidies was employed with a corresponding stick: By inflating gas prices to the point where ICEs become prohibitive, the Norwegian government forced most of the holdouts to make the switch to EVs.

Because of such government manipulations, Norway's booming EV market is a mirage. Christina Bu of Norway's Electric Vehicle Association admitted as much during an interview with *The Independent*: "The uptake has nothing to do with the Norwegian psyche or love of the environment, it came when people started to realize there were huge savings to be made."[25]

What makes Norway's example even more ironic is what actually makes the country's generous subsidies possible in the first place: oil. As David Yager put it, "Norway gets all its EV subsidy money by selling oil to the rest of the world."[26]

Norway is Europe's largest oil producer: Even with a downturn in production for 2018 and 2019, the Norwegian Petroleum Directorate anticipates an average output surpassing 4 million barrels per day by 2021.[27] In 2017, the value of Norway's hydrocarbon exports totaled NOK 442 billion ($50.5 billion)—fully half of all exports—and total net cash flow from the petroleum industry reached NOK 180 billion ($20.6 billion). Thanks to higher oil and gas prices, the Directorate predicted cash flow to hit NOK 286 billion ($32.7 billion) in 2019.[28]

"We have a very hypocritical policy," says Daniel Rees, an advisor on transportation and the environment for Norway's opposition Green Party.[29]

"What we have proven in Norway is that if you give enough subsidies and impose enough restrictions on fossil fuel vehicles, people will buy electric," Andreas Halse, the environmental spokesman in Oslo for the opposition Labour Party, told the *Financial Times*. "If we want to continue to be an example for the rest of the world we need to show how this can be commercial. We need to get there because we can't rely on public finances forever."[30]

Norway's leaders are talking about removing subsidies, but not until the mid-2020s or possibly even the 2030s. When—and if—that happens, it will be interesting to see how it affects the demand for EVs.

Cutting Subsidies, Encouraging Innovation

As Michael Lynch, a petroleum economics and energy analyst, commented in a *Forbes* article, "Subsidies and mandates are normally signs of inferior products and consumers are the ultimate judge, no matter the pontificating of experts, politicians, non-governmental organizations and inventors about the superiority of their favored solution."[31]

The government has tried to prevent supporting inferior products by mandating that funding only go to commercially sound enterprises: To receive funding through the ATVM program, companies must be "financially viable without the receipt of additional federal funding for the proposed project other than the ATVM loan."[32]

However, there's a plethora of capital in the private markets. This is especially true in the transportation sector . . . and even more so when fuel is involved. Fuel is a multi*trillion*-dollar global market, meaning any company that captures even a sliver of market share stands to make billions in profit. Global consumers are eager to find an affordable, clean, reliable source of alternative energy.

Financially viable EV companies don't need government assistance. There is plenty of demand for new tech. There is plenty of room for innovation. In fact, the transportation market is ripe for it.

CHAPTER 7

CHINA IN THE DRIVER'S SEAT

—OR NOT

China leads the world in everything from eggplant farming and steel production to mobile phone usage and IPOs. It's hardly surprising it plans to dominate the EV market, too.

By 2025, China wants 20 percent of all auto sales—about 7 million cars per year—to be plug-in hybrids or battery powered.[1] The hope is that adding EVs to the transportation mix will help China clear what is some of the world's dirtiest air: When it comes to overall carbon dioxide emissions, the nation ranks #1 in that category, too.[2]

Rushing to mine China's EV prospective gold rush are hundreds of car manufacturers. That includes big international names like GM, Ford, and Nissan and domestics such as BYD—an acronym for Build Your Dreams—which is backed in part by Warren Buffet.

With more choices becoming available, sales of new electric vehicles

(NEVs)—a category that includes plug-in hybrids, fuel cell electric cars, and battery-powered EVs—are clearly on the rise in China. In 2018, the total exceeded 1.03 million, an increase of 68 percent over the previous year, and of that number, a whopping 80 percent of them were battery EVs. In fact, one of every two EVs sold in the world in 2018 was purchased by a Chinese buyer; looking at it another way, there were more EVs sold in China than in all other countries combined. According to Bloomberg, in 2022, sales of passenger EVs in China will top 2.5 million. That's short of the 7 million figure the government has in mind for 2025, of course, but still a sign things are cruising along in the desired direction.

If you think this reflects some release of pent-up desire by Chinese consumers to go green, think again.

China's big automakers, including state-owned BJEV, are investing billions of dollars in electric buses and full-size EVs. The central government plainly has a vested interest in steering consumers toward EVs. And it has been doing so at the expense of gasoline-powered cars, making it difficult—if not downright impossible—to buy one.

ICE Under Fire

To achieve its EV goals, China is relying on a full slate of interventionist policies that remove consumer choice from the buying equation and force the hand of manufacturers who want to compete in the world's largest passenger car market. They include what has been called "the single most important piece of EV legislation in the world": the requirement that every car manufacturer operating in China, domestic or foreign, produce a certain number of NEVs each year, with the total increasing annually.

In addition, for nearly a decade, the government has provided liberal subsidies, including tax incentives, to manufacturers and consumers of full-size EVs. We're not talking about chump change here: Bloomberg reported that in 2017, China's central and local governments poured $7.7 billion into the subsidies.[3]

It's a universal truth that EVs cost more than gasoline cars. To entice a price-sensitive shopper to purchase a more expensive EV with a range longer than 250 miles, the government might toss in as much as $7,900. That would knock down the all-in cost of the typical full-size EV by about 30 percent.

For about a decade, China has experienced record-breaking passenger car sales growth. According to data from the China Association of Automobile Manufacturers (CAAM), when the China personal car boom took off in 2009, the country sold 10.3 million units—up considerably over 2008, when Chinese consumers bought 6.7 million cars.[4] By 2015, 21.2 million passenger cars were sold in China—outpacing the United States, long the world's dominant market, by nearly 4 million units. Although China's demand for gas-powered cars is said to have peaked in 2017, another 27 million new and used combustion engine cars rolled off Chinese lots in 2018.[5] That's nearly 27 times the number of EVs sold during the same period.

To encourage drivers to make the switch from ICE to EVs, a number of China's largest cities—Beijing, Shanghai, Guangzhou, Guiyang, Tianjin, Shijiazhuang, Hangzhou, and Shenzhen—have enacted strict policies that make it far more expensive to purchase a gas car and exceedingly difficult to operate or park one on certain streets.[6]

As if that weren't tough enough, throughout China, getting an ICE license has become such a torturous and expensive prospect that many would-be owners simply give up and go electric. And there's no attempt to hide the pro-EV bias: In Shanghai, for example, you need a $12,000 permit to buy an ICE car, but there's no such fee attached to EVs. In Beijing, since 2010, the only way to buy a new gas-powered car is to first win one of the city's bimonthly license plate lotteries. It can take as long as two years for a consumer's number to come up.

In Asia, Coal Is Still King

As part of their pro-EV playbook, lawmakers have even taken aim at overcoming one of the most serious roadblocks EVs face: range anxiety.

China's government has accelerated the construction of charging infra-structure, building 330,000 charging points at 70,000 stations across the country in hopes of developing enough charging infrastructure to support 5 million EVs by 2020.

The only thing is, those charging stations will likely be fueled by coal.

Given its pattern of firsts, it's probably not surprising to anyone that China produces more electricity than anyone else in the world. Some of it is generated by hydropower, but as much as two-thirds came from coal. And even though the country claimed it was committed to grow-ing the share of non–fossil fuels in its energy mix, China added 200 GW of coal-fired capacity in 2017.[7] Considering that Rice University's Baker Institute calculated that for each million plug-in electric passen-ger cars China puts on the road, it would likely create an additional 740,000 tons per year of coal demand, it's easy to see why they need the additional capacity.

Of course, burning coal is a dirty business. In terms of CO_2 emissions, the added coal burden is like operating 1 million BYD Fo gasoline-powered passenger sedans.

In other words, China is merely shifting its "pollution problems from a semi-dirty tailpipe to a potentially much more emissions-intensive power plant smokestack," the Baker Institute working paper on China and oil said. "In this sense, a pure plug-in electric car running on grid power in many parts of China (aside from areas where grid supply comes primarily from hydro or nuclear plants) is effectively still as carbon intensive as a fully gasoline-powered compact car, virtually negating the environmental benefits of going full electric."[8]

By the way, China isn't the only Asian country to be facing this sce-nario. In 2018, researchers at Malaysia's Penang Institute not only doubted the potential of EVs to cut that nation's greenhouse gases; they also sug-gested the vehicles could have a negative effect. The country's national power grid relies heavily on coal and fossil fuels. Introducing EVs into the market without creating a greener electricity source is akin to putting the car(t) before the horse, according to study author Darshan Joshi.

"A focus on EVs without first addressing the composition of the electricity grid would be premature and instead lead to a rise in the nation's transport sector emissions," Darshan said.

After calculating the carbon dioxide produced by a car per gram, per kilometer (g/km CO_2), Darshan concluded the ICE Perodua Bezza produced far less CO_2 than the other cars he studied: Tesla Model 3 S, Nissan Leaf, Tesla Model S 100D, and a Tesla Model X P100D.

"This evidence suggests that a shift to EVs in Malaysia today would be detrimental to climate change efforts. A more environmentally-friendly policy would instead incentivise fuel-efficient ICEVs, as well as HEVs, which are for the most part considerably less polluting than EVs," he said.[9]

In Vietnam, the situation is similarly fraught.

Ho Chi Minh City is considered the most polluted city in Southeast Asia and Hanoi isn't far behind. To try to make a dent in those dubious honors, in 2016, Vietnam spent nearly $500,000 on purchasing EVs from overseas markets. And EV sales are on the rise: By 2025, the Ministry of Industry and Trade expects they'll reach 600,000 units each year. But even that number is unlikely to offset the country's coal-generated electricity. Currently, Vietnam gets nearly half of its electricity from coal, and that figure is expected to rise to 56 percent by 2030. The trade-off between the tailpipe and the smokestack isn't even close.[10]

It's Not the Cars—It's the Factories

Despite science to the contrary, China is standing pat on its EV evangelism. And with its outsized impact on just about everything it touches, it's reasonable to wonder what kind of effect China's plan will have on the global economy.

After all, China's oil consumption rose steeply between 2009 and 2015, about the time the country's emerging middle class went on what seemed to be a boundless car-buying spree. In 2013, in fact, China became the first country to sell more than 20 million cars in a single year. For every million new cars sold since then, it's estimated that additional gasoline demand has grown an average of 11,000 barrels per day.

But the fact is, that's just not that much oil. Passenger vehicles create far less demand for oil than most people think. In 2016, for example, gasoline for all the cars and light trucks worldwide represented only about a quarter of global demand.

If it wasn't at the gas pump, then where did China become the world's second-largest oil consumer and top oil importer? Far and away, it was on the factory floor.

As *The Economist* reported, China makes half of the world's goods, including about "80 percent of air-conditioners, 70 percent of mobile phones, and 60 percent of shoes"[11]—and none of it gets done without petroleum. During the first quarter of 2019, as production rose and China's economy expanded 6.4 percent, the nation was even more oil thirsty than ever: Reuters reported that between January and April, China imported 164.9 million tons of crude, an increase of 8.9 percent over the same period in 2018.

With China's refining throughput expected to reach 12.7 million barrels per day in 2019, up 600,000 from 2018, demand for petroleum should continue to be strong—far too high to be displaced to any degree even by a surge in EV sales.

Oil Is Here to Stay

To date, EVs' impact on the oil market remains relatively small: Bloomberg says 1,000 battery electric vehicles remove just 15 barrels of oil demand per day. By 2040, the group expects that, worldwide, EVs could displace as many as 6.4 million barrels a day of oil demand—but even that is less than the effect of fuel efficiency improvements to gasoline-powered cars.[12]

Economist Fatih Birol, executive director of the International Energy Agency, is similarly unmoved by the potential effect of electric cars on the oil economy.

"To say that electric cars are the end of oil is definitely misleading," he told a panel at the 2019 World Economic Forum in Davos. "This year we expect global oil demand to increase by 1.3 million barrels per day.

The effect of 5 million cars is [to diminish that demand by] 50,000 barrels per day. 50,000 versus 1.3m barrels."[13]

Researchers at Columbia University agreed that a decline in passenger vehicle oil demand does not necessarily mean that total oil demand will decline.

"There may be strong-enough growth in petrochemicals, aviation, and trucking to offset any decline in the passenger vehicle sector," Marianne Kah, author of "Electric Vehicles and Their Impact on Oil Demand: Why Forecasts Differ," said.[14]

The End of Subsidies, the End of the Line?

The wild card in China's EV future and the consequences on global energy markets is this: Many of the government's subsidies are lapsing. In other countries, this has had a chilling effect on EV manufacturing and consumer adoption.

Beijing announced in March 2019 that it is cutting back on subsidies. Gone are nearly limitless levels of support, like the $590 million subsidy that helped BYD transition from a battery maker into an EV giant.[15] The government is doing away with incentives for EVs with a range less than 155 miles; higher-spec EVs may see subsidies reduced by as much as 60 percent.

China claims that scaling back subsidies will force domestic EV makers to be more innovative and agile in developing new technologies. But we can't overlook the economic reality of "free" money or the desire by China's government to trim expenses. As Jack Perkowski wrote for *Forbes*, "Under the current subsidy program, subsidy payments would rise to approximately $20 billion in 2020 and $70 billion in 2025. In order to put this number into perspective, the annual budget of the Chinese government was RMB 20.3 trillion ($3.1 trillion) in 2017, and the government ran a fiscal deficit of RMB 3.1 trillion ($460 billion)."[16]

With China's EV companies already turning a profit, Perkowski added, it seemed reasonable for the government to shift the burden of subsidizing the development of the industry from the government to the automakers themselves.

The move is likely to separate the winners from the losers among China's 400-plus EV manufacturers.

"With the subsidy adjustments, some less technologically advanced EV startups will disappear," Zhou Lei, a Tokyo-based partner for Deloitte Tohmatsu Consulting, told Bloomberg News. "There will be a reshuffle."[17]

And even though the new rules didn't go into effect until June 2019, at least one manufacturer experienced their consequences: BYD admitted slow sales during the first quarter of the year, and their stock fell on that news.

It's hard to say what effect subsidy slashing will have on consumers. Most experts believe that the EV makers won't raise prices. China's EV sales were close to 1.18 million in 2019, according to InsideEVs, falling short of CAAM's prediction of 1.6 million by year-end.[18]

By contrast, the experts at the Baker Institute are convinced that these changes will dampen consumers' force-fed interest in EVs. They see the issue as this: With EVs, "expenses, inconvenience, range limitations, and the costs of chargers and other supporting systems fall on the individual owner whereas the costs of liquid-fuel infrastructure like gas stations are distributed across society because so many people use them. As such, China—like other major global markets—is most likely to continue using gasoline propulsion for passenger vehicles."[19]

For a nation accustomed to being in the driver's seat in so many categories—from sowing eggplants to seeding public offerings—China may not achieve the EV leadership it aspires to, at least on the timetable it has set out. But with its crude imports increasing to 10.1 million barrels per day in 2019, according to EIA, and refining output on the rise, China will continue to prop up the global oil industry, regardless of EV uptake.

WHAT HAPPENS WHEN GIMMES ARE GONE?

China's plan to cut back incentives may put the chill on EVs and heat up the ICE market instead. It's happened before, even regionally.

Under the Thai government's "Electric Vehicle Promotion Plan for Thailand," for example, Thailand nearly doubled the number of hybrid passenger cars on its streets between 2014 and 2018, from 60,000 to more than 102,000. During the same period, it added some 1,400 battery electric vehicles.[20]

Behind the upswing is a tax-exemption program that rewards EV manufacturers and suppliers by reducing corporate and import taxes. The biggest breaks go to makers of "pure" electric cars, although no one has taken the bait yet. As one of the world's top producers of EV batteries, Thailand is giving an additional boost to car makers that use locally built batteries; they'll receive a reduction in excise taxes. At least eight international automakers have applied for the hybrid incentives; so far, only Toyota has received them.

Still, experts think that HEVs and PHEVs will command a 10 percent share of Thailand's auto market by 2025. To see that forecast through, the Thai government is also supporting the construction of public charging stations, to the tune of $1.3 million.[21]

But there are no buyer incentives. And without them, the outlook for 100 percent EVs, which come with hefty price tags, remains as murky as a Bangkok thoroughfare during rush hour.

Similarly, for nearly two decades, all newly registered ICE vehicles in Hong Kong have been subject to significant taxes—so significant that they can sometimes double the cost of the car. Meanwhile, these taxes were entirely waived for EVs (see figure 7.1).

But in April 2017, that all changed: The tax waiver was capped at $12,500. As a result, the average Tesla Model S 60 price shot up from $72,900 before the new tax policy to $118,400. The consumer response was immediate: According to Echo Huang's article on Quartz, "Not a single newly purchased Tesla model was registered in April, according to data from Hong Kong's transport department."[22]

continued

Vehicle	Sticker Price (HK$)	Tax (HK$)	After-tax Price (HK$)
Mercedes A 180 FL	$314,000	$186,500	$500,500
Mercedes E 200 Premium Edition	$564,000	$506,100	$1,009,500
Tesla Model S 70D	$619,000	$0	$619,000

Figure 7.1. *Cost comparison for ICEs and EVs in Hong Kong.*
Source: https://qz.com/559226/in-hong-kongs-luxury-car-market-a-tesla-is-cheap/.

Since then, Hong Kong has walked back its decision in hopes of resuscitating the market. In February 2018, the government announced it is adding a new first registration tax (FRT) waiver that will save buyers as much as $32,000.

There's a catch, however: Drivers can claim the waiver only if they "scrap and de-register their own eligible old private car (private car with an internal combustion engine or electric private car) and then first register a new electric private car."

Although the new policy may drive some EV sales, it's unlikely they will reach what Electrek called the "artificially high demand" that occurred as buyers rushed to take advantage of the initial tax policy.[23]

OTHER ELECTRICS IN JEOPARDY

In addition to taking aim at ICE, local and central governments have also drawn a bull's-eye on two mainstays of China's lower-end transportation market: the e-bike and micro-EV.

Electrified transportation is not a just-out-of-the-gate phenomenon in China: There are at least 200 million electric-powered two-wheelers on the streets, with another 15 million added every year. Because they top out at less than 12 mph, they are considered "bicycles," meaning they can be operated in bike lanes along with their pedal-powered alternatives.

Of course, these e-bikes, as they're more commonly known, might not be what come to mind when most people think of EVs: They hardly have the cache, cargo space, or seating of a Tesla—or the price tag, for that matter. Buying a typical e-bike sets the Chinese consumer back as little as $200 on China's e-commerce site Alibaba. For lower income consumers, e-bikes are the new family cars: While it's tough to imagine Americans considering an e-bike as a substitute for an SUV, in China it's not uncommon to see Mom, Pop, and the kids all crammed onto one, narrow seat.

What e-bikes lack in roominess, they make up in environmental friendliness, at least at first glance: Experts suggest that because e-bikes use so little electricity—about the equivalent of getting 1,000 miles per gallon[24]—they're pretty clean, even though many are equipped with pollution-producing lead-acid batteries (e-bikes are a leader in lead use) and all run on coal-fired electricity.

Yet some Chinese cities have restricted their use or banned them altogether. Although most claim safety concerns—China's Ministry of Industry and Information Technology said that between 2013 and 2017, e-bikes caused more than 56,000 traffic accidents, resulting in more than 63,000 injuries and 8,000 fatalities[25]—that's just part of the equation. Consider that in Shenzhen, not so coincidentally the home of upstart automaker BYD Company, police confiscated several thousand e-bikes and detained hundreds of users.

continued

http://www.onsetbayphoto.com/blog2/?p=610;
https://www.quora.com/Why-are-motorcycles-banned-in-China.

Owners of micro-EVs have fared only slightly better.

Calling them "a quirky subplot in China's push to become a world leader in electric cars," *The Wall Street Journal* (*WSJ*) reported in 2018 that Chinese manufacturers are dipping into the lower income market with low-speed four-wheelers.[26] These souped-up Cozy Coupes are small, cheap, and slow: Although they come close to Smart car dimensions, they sell for about one-fifteenth as much and amble along at up to 43 mph—about half the top speed of a Smart car. According to the *WSJ*, Chinese buyers purchased about 1.74 million micro-EVs in 2017. While they are especially beloved in the rural provinces, there's demand even in big cities, largely because micro-EVs are easy to park and can weave through traffic jams. And, despite their speed profile, in many areas they can use both car and bike lanes.

As far as their green profile, micro-EVs come close to e-bikes, producing fewer tailpipe emissions but relying on dirty lead-acid batteries and electricity that comes from coal. But that's not where the similarity ends: As with e-bikes, the government isn't content to leave the market alone. Efforts to regulate micro-EVs' weight and size limits, require crash protection, and mandate the use of lithium-ion batteries could kill off the vehicles. Many cities are already banning the pint-sized roadsters, including some near Gaotang, a micro-EV factory town that is home to more than a dozen manufacturers.

https://www.wsj.com/articles/chinas-giant-market-for-tiny-cars-1537538585.

PART II

THE CARS

EV OPTIONS AND OWNERSHIP

CHAPTER 8

EV AUTOMAKERS:

MAKING IT OR BREAKING IT

When Aptera Motors released its three-wheel EV prototype, it seemed to be ushering in a new era of ultramodern, supercool EV transportation. "At first glance the Aptera Typ-1e and the Typ-1h (the hybrid version which is almost identical in appearance to its all-electric stablemate) screams one word—AERODYNAMICS," Noel McKeegan of *New Atlas* wrote in 2007. "This is no coincidence as the futuristic design was evolved entirely on the premise of finding the optimum shape for a two-passenger vehicle."[1]

But despite its auspicious beginnings, California-based Aptera folded in 2011, just five years after it rolled out the Typ-1e and Typ-1h. The vehicles' futuristic "cool factor" simply wasn't enough to keep it afloat.

Aptera's CEO Paul Wilbur and former marketing VP Marques McCammon told Green Car Reports in 2011 that the company would have survived if it had managed to get funding in time from the U.S. Department of Energy (DOE) through its Advanced Technology Vehicles

Manufacturing Loan Program. "We should have raised the money ourselves rather than relying on the DOE,"[2] Wilbur said.

Of course, Aptera wasn't the only EV startup to go under. Another California-based company, Coda, made a four-door sedan with an 88-mile range and a top speed of 85 mph. The company was in business for one year and sold 177 cars before declaring bankruptcy.

The reason for Coda's demise? For one thing, their sedan wasn't much to look at. "Coda was not producing great-looking cars," Chris Woodyard of *USA Today* wrote in 2012,[3] shortly after the company declared bankruptcy. "They resemble an attempt to update the econobox designs of the 1980s or 1990s. They were brought in from China, and the powertrains were installed in California."

The Coda didn't offer much in the way of performance and owner experience, either. All 177 cars the company sold were quickly recalled for issues with their side curtain airbags. There was zero "cool factor." The manufacturing left much to be desired. "They didn't drive particularly well, either, or feel like they were of high quality," wrote Woodyard.

In a sea of sleek and futuristic designs, the Coda simply didn't bring anything new to the table. "If buyers were snapping up electric cars, as some thought they would be by now, none of that would have mattered," Woodyard wrote. "Coda would have just blended in. But with electric-car makers struggling, Coda was easy to shirk off."

The takeaway from both companies' experiences is this: It is very, very difficult to successfully manufacture EVs. In addition to contending with the typical challenges of any new business venture—such as raising capital—EV manufacturers are also hindered by the hurdles ranging from limited charging infrastructure to long charging times to a still-skeptical consumer base. Michael Lynch, president of Strategic Energy and Economic Research, Inc., calls these issues an "inconvenience penalty." And EV manufacturers are definitely being penalized by lack of sales: According to a 2017 survey by McKinsey & Co., 30 percent of U.S.-based respondents said they'd considered buying an EV, but only 3 percent of them actually did.[4]

"[EV] advocates claim the problem is exaggerated, but reading of real-world experiences suggest otherwise," Lynch wrote in a 2016 guest article for *Forbes*. "Even positive articles include discussion of the amount of advance planning needed to ensure finding charging stations or rationalizations for the inconvenience. Telling customers that your product isn't that bad, or that its shortcomings can be overcome without too much trouble is not a winning strategy."[5]

That's not to say EV makers are doomed. It just means that the manufacturers can't assume large portions of the buying public are clamoring to buy EVs. EV makers still need to offer value, based on what *consumers say they need and want*.

Tesla: Doing EVs (Mostly) the Right Way

While Aptera and Coda (and, of course, Fisker!) can be viewed as cautionary tales for any company interested in producing EVs, Tesla is a case study of what can happen when EVs are done right. The company has taken a proactive approach to identifying, and addressing, its customers' needs.

Tesla's path has not been without potholes, and its early use of government subsidies is problematic. But the company's success is undeniable. "The compound annual growth rate of Tesla's gross profit over the last five years is 91.8 percent," writes Trent Eady of *Seeking Alpha*. "That's faster than Facebook and Amazon combined. Tesla is one of the fasting growing companies in the world, and possibly the single fastest growing manufacturing company."[6]

Tesla's ambition is simple: The company wants to change the world. It was founded with the goal of nudging the world toward a sustainable transportation revolution, and it set out to do that by proving that EVs can be better, more fun to drive, and faster than ICE vehicles. Tesla is so committed to its objectives that the company made its technology open source in 2014 so other companies can build EVs of their own. "Most of the good that Tesla will accomplish is by cutting a path through the jungle to show what can be done with electric cars," CEO Elon Musk said in 2015.[7]

While Tesla has missed many of its self-imposed deadlines over the years, including targets for production and delivery, the company has moved steadily toward meeting its core objectives. In 2008, Tesla launched the Roadster, the first serial-production, all-electric car to use lithium-ion battery cells. The two-door sports car had a range of 244 miles on a full charge. It went from 0 to 60 mph in 4.4 seconds and covered a quarter mile in 13.3 seconds at 104 mph. But at a starting price of $110,000, the Roadster wasn't a car for everybody. Tesla sold about 2,400 of them.[8]

Tesla's next car, the Model S sedan, received rave reviews for performance and design when it was released in 2012. "There's no question that the Model S is a desirable automobile. It's attractive, feels well-built and luxurious, offers features that no other EV does, handles the way a serious sports sedan should, and generally comes off as a wholly considered proposition," Davey Johnson of *Autoweek* wrote at the time.[9] The Model S, which sold for $57,490, offered three battery options with estimated ranges of 235 to 300 miles. The battery option with the highest performance gave an acceleration of 0 to 60 mph in about four seconds and a top speed of 130 mph.

Tesla's EV selection has continued to grow. In September 2015, the company released its Model X, an all-electric SUV with seating for up to seven adults, all-wheel drive, and up to 295 miles of range per charge. Tesla describes it as the quickest SUV in production, able to accelerate from 0 to 60 mph in 2.9 seconds.

And like the Model S, the $80,000 vehicle received mostly glowing reviews. Reviewers may have ridiculed the Model X for its falcon-wing doors, but they were in awe of the SUV's $10,000 "Ludicrous Speed" option.

"That special fuse increases the battery's output to 1500 amps (up from 1300), and the available output rises to 532 horsepower," reviewer Tony Quiroga of *Car and Driver* wrote about the 2016 model. "With or without Ludicrous Speed, the full 713 pound-feet of torque is available with every punch of the accelerator below 50 mph. That neck-straining torque certainly gives the sensation of 700 horsepower. Or of falling off a tall building."[10]

Tesla's most affordable offering yet, the Model 3, offers 220 miles of range at a price point in the mid-$30,000s. And while it has been described as a "stripped-down" Model S, the Model 3 still packs a punch: The car goes 0 to 60 mph in six seconds. Missed deadlines didn't seem to dampen consumer interest. Tesla by late 2017 had received 500,000 pre-orders and enjoyed record sales in 2019, selling more cars than in the previous two years combined, according to The Verge.[11]

Tesla's growing value is undeniable: The EV manufacturer's net worth is over $150 billion as of May 2020.[12] Perhaps the fact that Tesla has taken a long-term view to maximizing shareholder profit, and is working to completely change transportation, helps this company thrive while others have failed.

Tesla has gone to dramatic lengths to motivate consumers to embrace EVs, even when that calls for creating new technologies and infrastructure. The company spent $1.98 billion to construct its 5.8 million-square-foot Gigafactory in Nevada, for example, which provides the scale it needs to mass-produce the high-performing lithium-ion batteries that are essential if the company wants to meet its stated goal of producing over a million in 2020.[13]

But, maybe more important, Tesla has been building charging stations since 2012. By the end of 2019, it had about 34,000 charging stations worldwide. About 13,000 of those are "Superchargers" that need only about an hour to fully charge a Tesla battery. The remaining 21,000 "destination chargers" take a bit longer to charge and are targeted for long mall visits or hotel overnights.[14]

Fred Lambert of Electrek noted that Tesla's Supercharger plans eclipse anything done by third-party networks so far. For example, "ChargePoint said that it had only 371 DC fast-chargers in the United States last year. Tesla will have more in Los Angeles' metro area alone by the end of the year."[15]

It's worth noting that not all major automakers are committing to an electric future. Sergio Marchionne, CEO of Fiat Chrysler Automobiles, told a crowd of reporters in October 2017 that the world still lacks a viable economic model for electric cars. Even Tesla, he says, has failed on that front. "As much as I like Elon Musk, and he's a good friend and actually he's done a phenomenal job of marketing Tesla, I remain unconvinced of [the] economic viability of the model that he's pitching," Marchionne said. "Until the cost of batteries and technologies associated with electric cars drops," electric cars will remain in limited use.[16]

On the Market: Established Manufacturers and Newcomers

While Tesla tends to get the most attention for its EVs, it's far from the only company manufacturing them. Between 2016 and 2018, major automakers throughout North America, Europe, and Asia announced plans to begin or ramp up their EV production. One of the most dramatic of these announcements took place at the 2018 North American International Auto Show in Detroit, when Ford Motor Co. said it would spend $11 billion on EVs in the next five years with the goal of releasing 24 hybrid and 16 fully electric vehicles by 2022.[17]

Mercedes-Benz, Jaguar Land Rover, and Volvo say they'll electrify their entire portfolio, too, and have set deadlines for themselves between 2019 and 2022. Other companies have announced partnerships to bolster EV development: Toyota and Mazda are collaborating with Denso, Toyota's biggest supplier, to make affordable EVs, including cars, SUVs, and light trucks. And Renault, Nissan, and Mitsubishi have formed an alliance to develop new systems for their vehicles, with an emphasis on EV enhancement and development.

General Motors, meanwhile, has announced plans to release 20 all-electric vehicles by 2023, eventually phasing out ICE vehicles altogether. And the company is off to a promising start: GM's current Chevrolet fleet includes a trio of forward-thinking offerings: the Volt, a plug-in hybrid sedan; the Bolt, a sporty electric hatchback; and the Hybrid Malibu, which was released in 2017. Of the three, the Chevy Bolt, which offers 240 miles of range and a relatively affordable base price of $37,500, has been a favorite of consumers and critics: In 2017, *Consumer Reports* named the Bolt Chevy's most reliable car; it also won *Motor Trend*'s "Car of the Year" award.

Just how popular is the Bolt? Chevy sold 23,297 of the vehicles in 2017, putting the model in second place, sales-wise, after the Tesla Model

S, which saw total sales of 27,060 the same year.[18] The top five EV models in terms of sales in 2017, after the Model S and Chevy Bolt, included Tesla's Model X, the Nissan Leaf, and the Ford Fusion Energi.

But tried-and-true brands like Chevy, Mazda, and Mercedes aren't the only ones gunning for Tesla's spot on the top of the EV heap. Some are going with the proven Tesla combination of fast, sexy, and powerful vehicles. Those qualities were what Maryland-based Genovation Cars was going for when it converted a C7 Corvette into an all-electric supercar with 800 horsepower. The resulting creation, the GXE, is said to reach speeds up to 220 mph and go from 0 to 60 mph in less than three seconds. Of course, that speed and power come at a cost: The GXE is priced at $750,000.[19]

It would be fair to say Lucid Motors will be putting out cool cars, too. The Silicon Valley startup has said its upcoming EV, the Air, will cost about $60,000. The vehicle, which has reached a speed of 217 during testing, will have 240 miles of range. Lucid also plans to offer more expensive options with 315-mile and 400-mile ranges. Production is now expected to begin toward the end of 2020, and a $1 billion investment from the Saudi Arabian Public Investment Fund that was just completed in April 2019 should make this a reality.[20]

Then there are companies making a name for themselves as producers of EV hypercars—the upper echelon of the fast, luxurious vehicles that are considered supercars.[21] Croatian EV startup Rimac Automobili was described as a pioneer in EV hypercars when it introduced the $1 million Concept One. Company founder Mate Rimac engineered this car with separate electric motors to power each wheel and linked them together with software. The car goes from 0 to 185 mph in 14.2 seconds. "That is insane. Stretch this car's legs on the public road and you're probably going to jail,"[22] wrote Nick Hall of Dgit.com.

Some are aiming to impress with more than the driving experience. EV manufacturers like China-based Byton are going all out with "infotainment" components. Byton's concept vehicle for its first product launch features Amazon's voice assistant Alexa, facial recognition access, real-time monitoring of health vital signs, hand-gesture control, and cloud-based data storage. With a 49-inch edge-to-edge dashboard display, an eight-inch tablet in the steering wheel, and screens on the back of the front headrests, Byton calls the vehicle a "digital living room on wheels."[23] It will go into production for sale initially in China, then branch out into American and European markets. Byton plans to price it in the mid-$40,000 range.

Byton is one of many China-based EV makers that may soon build a presence in North America and Europe. Another was founded in 2014 by Henry Xia, who was working in research and development for the Guangzhou Automobile Group when Tesla announced it was opening its patents. Recognizing the opportunity to launch a startup, Xia created Xiaopeng Motors. In 2017, the young company introduced a beta version of the Xpeng, an all-electric SUV with a range of 186 miles.[24] Since then, Chinese e-commerce giant Alibaba invested in Xiaopeng and now has a 10 percent stake in the startup.

Some of the newcomers to the EV scene aren't from the automotive world at all. Home appliance maker Dyson is a prime example: The company announced plans to release its own EV by 2020 without partnering with an automaker. "We want to do our own thing; we want to do it our own way," CEO Max Conze told *Wired*. "We want to do a Dyson car the way that we think it needs to be done, and that requires us to have the engineering that can do the car end-to-end and also to own the manufacturing."[25] But the company killed the project in October 2019, announcing on its website, "We simply cannot make it commercially viable."

Will one of these newcomers one day bump Elon Musk from his position as the reigning king of EV technology? Maybe one day in the future we'll tell stories of how Henry Xia changed EV technology forever.

Perhaps the next big thing in EVs is yet to come. It's much too early to predict a winner. But it'll be a lot of fun watching the game.

CHAPTER 9

WHAT CONSUMERS
DON'T LIKE

(OR THINK THEY DON'T)

At the beginning, Bradley Niemcek said in a personal interview, there
was a very big learning curve.

"For the first several weeks, I honestly had to study the manual . . .
out loud! I've never had to do that with a car before," he said. "And a big
complaint is that the owner's manual is a 'Chevy Group' car manual with
references to five other models. With all the complications for an electric car with complex systems, you'd think at the very least they'd have a
manual specific to the car. I still can't find how to change certain things."

Niemcek is not only a former race car driver, but he was also a member of the American Racing Press Association, even serving as secretary
for a term in the 1970s. With his press credentials, he had sideline access
to some of the automotive newsmakers of the day. From that perspective,

Niemcek has a solid understanding of cars and a sizeable grasp of mechanical engineering.

Despite his extensive knowledge, though, Niemcek says that his first long trip in his Volt was "a little scary." He compared the experience to flying an airplane: "Between the gas engine plus the battery motor plus all the mind-boggling electronics, I was thoroughly mesmerized by all the stuff I had to take care of."

If a seasoned race car driver and auto enthusiast like Niemcek had a longer-than-expected adjustment period to his 2016 Chevy Volt, one can only imagine how the average American consumer would fare. The steep learning curve and "mind-boggling electronics" could potentially push away drivers who don't want to have to teach themselves how to drive—again.

Even the Professionals Are Novices

Early adopters aside, many of us tend to take a wait-and-see attitude when it comes to new technology—and EVs are no exception. While waiting for us to all jump on its bandwagon, the EV has fallen prey to a combination of myths, false assumptions, and the lack of a standardized, reliable sales process.

When it comes to selling and servicing EVs, "green" takes on a whole new meaning. Despite the near-decade-long existence on the market, car dealerships have been, for the most part, very slow on the uptake: Models either aren't at dealerships, or they're in limited supply, or they're only sold in certain target markets. Individual salespeople and service technicians have limited exposure and lack the expertise and understanding to answer questions or give fact-based recommendations.

EV drivers find themselves much better educated about EV technology than the team at the dealership and the local repair shop. A prime example is the standard oil change: Industry recommendations for ICE cars, which typically fall in a three- to six-month window, are still the fallback—even though PHEVs don't burn oil like the ICE. As a result, less-informed EV drivers have fallen into the trap of unnecessary maintenance.

"I've put on 15,000 miles in one year of ownership and only had the oil changed because the dealership told me I had to! I didn't agree with them—my car said there was 85 percent life left in the oil. But the service guy said that oil breaks down over time as well as over use, yadda yadda. So I caved," said PHEV owner Joshua Feyen in a personal interview.

But unnecessary oil changes aren't the only issue: Two months after Feyen purchased his EV, it was rear-ended. The vehicle sat in the dealership for a solid month before the technicians could fix it.

Niemcek says he experienced similar issues trying to find service for his Chevy: "I realized this chilling fact: A lot of dealers didn't want anything to do with the Volt. They won't service them. They'd have to train special staff, order special tools."

There are, however, forward-thinking service shops that seem to understand the need to broaden their horizons. Feyen pointed out, for example, a local mechanic whose signage explicitly states, "We Repair Electric and Hybrid Cars!" and cited a recent radio show about a mechanic who sees his shop going out of business in ten years, so he's educating himself on EV technology.

What Holds Potential Buyers Back?

According to John Voelcker, editor of Green Car Reports, "Plug-in electric cars are unfamiliar to most buyers, and require more explanation—sometimes a great deal more—along with specific types of support during and after the sale."[1]

Unfortunately, many potential EV buyers find very little support during the decision-making process. For example, after months of research, Niemcek finally decided on his EV of choice—then ran into a conundrum. "EVs in general were hard to find around here. For a while, I would have had to drive to a dealership 300 miles away just to see one in-person for a test drive."

Both sellers and buyers are likely unprepared for EVs. In 2017, market research firm Ipsos RDA released a study stating that U.S. car dealerships aren't doing an especially good job of selling EVs in the first place. The firm sent mystery shoppers to 141 dealerships in the country's ten largest EV markets. With the exception of Tesla, the dealerships did not impress. The mystery shoppers reported that dealerships failed to customize their sales process for EVs, and, in many cases, EVs weren't even on the lot for customers to view or test-drive. Finding salespeople knowledgeable in EVs, or even salespeople who had brochures available, was hit-or-miss from dealership to dealership.

"The biggest takeaway was that there was just a lot of inconsistency," said Mike VanNieuwkuyk, senior vice president of Ipsos RDA, in an interview with Greentech Media. "There aren't standards or processes that are implemented and communicated across the brand."[2]

In another worrisome pattern, salespeople who couldn't speak knowledgeably about EVs seemed eager to return to their comfort zones: Many tried to steer the mystery shoppers to ICE options instead of EVs. "This lack of support for the EV shopper lessens the likelihood that they will make the decision to go electric,"[3] said Todd Markusic, vice president of research at Ipsos RDA, in a statement.

Voelcker says this is why it's essential for consumers to do their homework before they start shopping: "You need to be better informed than the salesperson—who may not have the information, or may be incorrect."[4]

Enter the incorrect information—the myths, misconceptions, and general confusion—about the electric car.

MYTH: EVS HAVE NO PICK-UP

"I hear this a lot, and I have no idea why," Niemcek said. "Acceleration is quite impressive." This from a former race car driver! And posts to consumer forums on the Chevrolet Bolt concur: Drivers are almost universally pleased with the acceleration power of their EVs. In fact, common descriptors are "zippy" and "peppy."[5]

"The pick-up was surprising at first. Now I love squealing the tires for someone in the passenger seat," Feyen admitted.

What gives EVs so much zip? It's all about torque. An electric motor's full torque is there immediately from the start, whereas a gasoline engine needs higher rpms for more torque. *ExtremeTech's* Bill Howard wrote about "the incredible power potential when you combine an electric motor with a turbocharged gasoline (eventually diesel, too) engine. An electric motor develops maximum torque, or power, at 0 rpm. That's where a gasoline engine is weakest, and weaker still with a turbocharger that needs a half-second (from when you tromp the throttle) to spool up and boost the power."[6]

Perhaps this myth has its roots in truth, though: Torque does begin to level out at higher speeds. And the Nissan Leaf might have perpetuated this concern by introducing its 2011 model that came equipped with a "passing gear,"[7] which could have caused critics to picture a windup toy that sputters at higher speeds.

It is worth noting, regardless of the pick-up, EVs are generally known to have "iffy traction." The tires are designed to decrease rolling resistance to extend battery range as much as possible. This design means that they have sacrificed some of their grip on the road. The tires squeal easily and don't provide much traction on wet surfaces—not to mention snow or ice.

"It's not a car you feel like cornering quickly," Niemcek said. "From a handling point of view, it doesn't have a German car highway feel. There's something solid about the Jetta on the road; the EV just has a softer feel."

SOUNDS OF SILENCE

"Something I thought I'd really miss has become a plus, maybe the biggest plus for me. Sure, it's great to cruise past the gas station. But really, the silence is my favorite feature," Niemcek said. "Driving an electric car is so quiet, so calming. You can chirp the tires and rocket away from a stop sign, and it doesn't make any noise! As a former race car driver, I really thought I'd miss the sound of the engine. But I don't—not at all."

Remember when luxury ICE cars began marketing their soundproof interior? Well, that was basically thanks to heavy-duty insulation that muffled the sounds from outside. In contrast, EVs themselves emit

continued

very little noise to begin with. Environmental advocates also cheer the fact that these cars don't add to the prevalent noise pollution.

As an interesting sidenote, however, the silence has become a bit of a problem for some. Concerns over EVs' "dangerously quiet" operations have forced the National Highway Traffic Safety Administration to issue its late 2016 federal safety standard requiring new EVs to emit an audible sound when traveling at less than 19 mph.[8] (The concern is that people—especially the hearing impaired—may not hear them coming.)

MYTH: EV DRIVERS WILL BE STRANDED.

Again, there's truth behind this so-called range anxiety. Drivers of all-electrics who push the limits of their battery range will eventually reach a point where they'll need to stop for a charge.

The reality, however, is that this is an unlikely scenario. The U.S. Department of Transportation says that American drivers average 40 miles per day.[9] Most new EVs (even the "lower-end" models) can go 80 miles on one charge, and PHEVs can go at least 300 miles on a combination of electricity and gasoline; the range-extending engine switches on after the battery is depleted.

The technology already all but negates this anxiety, and rapid advances are only improving the situation.

Consider, for one, Tesla's April 2019 announcement that its new "long range" lines of Model S and Model X will travel 370 miles and 325 miles, respectively, on a single charge. Thanks to upgrades in the drivetrain and suspension, these models are showing a 10 percent improvement in range[10]—and it didn't really take a very long time in R&D to finalize these upgrades.

According to one Chevrolet Bolt owner, posting on a consumer review online forum, the fear of being stranded without a charge is outdated: "In a month of ownership, with a daily round trip commute of 30 miles plus weekend trips of 60+ miles, I don't think the battery ever dropped below half charge, and I don't charge every day. The 110V connector meets my needs well—no need for me to pay for a 220V charging station (110V charges about 4 mi/hr, times 10 hrs a night, more than meeting daily needs). I'm also averaging about 4.5 mi/kWHr, or 270 miles on a full battery."[11]

But the downside to longer-range batteries is the latest cause for panic among EV drivers, namely "charging time trauma." Eric Taub of *The New York Times* explained it this way: "Compared with a five-minute pit stop at your local gas station, charging an electric vehicle is a glacially slow experience. Modern electric cars still often need an entire night to recharge at home, and even at a commercial fast charging station, a fill-up can take an hour or more."[12]

John Heyward, professor of mechanical engineering at the Massachusetts Institute of Technology, summed up charging woes eloquently. "Holding a gas nozzle [today], you can transfer 10 megawatts of energy in five minutes," adding that recharging a Tesla at that rate would require "a cable you couldn't hold."[13]

Unpredictable battery range has also been cited. "Don't believe the range rating on the window sticker," one review of the Chevrolet Bolt cautioned. "Usually the warranty info in fine print will give you more realistic information about long-term use. Take that into consideration for your needs. Also, note that the rated range on the sticker assumes you are driving alone, not using the heat or the air conditioning, and not carrying any load."[14]

And those are some hefty assumptions.

There are, in fact, lots of external factors that can affect battery performance and range per charge. Hauling other passengers or gear adds weight; the more weight, the more electricity needed to propel the vehicle. Chilly outside temperatures impede battery range, as does turning on the heat inside the vehicle. (It's interesting to note, however, that one driver in Ontario, Canada, video-blogged a drive in his Chevy Volt on an insanely cold [–43.5 degrees Fahrenheit—and yes, that's *negative* 43.5 degrees!] morning.[15] Even in that extreme cold, with the heat on full-blast and the heated seats set to high, the driver only lost about three miles of predicted range.)

Terrain is also a factor. Driving uphill, for instance, burns lots of electricity, while driving downhill will actually boost the range, thanks to the regenerative braking capabilities. And highway driving—especially at speeds of above 55 mph—drains the battery faster than stop-and-go city driving (again, it's all about the regenerative braking).

MYTH: THERE'S NOWHERE TO CHARGE EVS

Today's EVs can be charged at any ordinary household electrical outlet, and public charging stations are beginning to pop up in shopping centers, auto service stations, and interstate rest stops around the country. Analysts predict that publicly accessible charging options will mushroom, out of necessity and consumer demand, as more EVs hit the road.

ChargePoint, the largest installer of vehicle charging stations in the United States, cites statistics that 80 percent of vehicle charging is done at home. Residential chargers are typically one of two levels. Level 1 is an ordinary 120-volt electrical outlet, like any basic household power outlet, that provides 1 to 1.5 kilowatts of electricity. Level 2 is a step up, delivering 7 to 9 kilowatts via a professionally installed charger connected to a 240-volt AC outlet, the kind used by some large household appliances. To put these into perspective, it would take about 30 hours to completely charge the 115-mile battery on an electric Ford Focus, using a level 1 charger. A level 2 charger would cut that time down to about five and a half hours.

"I've considered installing a fast-charge station at home, but my lifestyle isn't that demanding to really need one," Niemcek said. "Especially since I read something that the difference is negligible, so there'd be a long payback."

ChargePoint has unveiled plans to install many new fast-charging stations in the United States in 2018, but it is unclear how many will be installed or where these will be found. Meanwhile, the Volkswagen subsidiary Electrify America—established as part of its settlement for falsifying diesel vehicle emissions results—is investing $2 billion in charging infrastructure and EV education nationwide. This is great news . . . for EV drivers in California, where 40 percent of this funding is earmarked.[16] But the rest of us might still be a bit stretched. Plus, the target implementation date isn't for a few years.

So for now, despite forward strides, this is one perception that's not really a myth—at least not in the United States. As of spring 2019, the Department of Energy's Alternative Fuels Data Center counted more than 21,000 charging stations for a total of 63,000-plus charging outlets of varying charging speeds.[17] At first blush, that sounds impressive . . . until we compare that to the approximately 122,000 gas stations nationwide, many of which have ten or more pumps.[18]

True, this "charging station argument" may become less important and have less impact on the decision to buy an EV as technology pushes battery range high enough to get most people where they want to go on one charge. But at present, EV drivers notice that public charging is still something of an overpriced novelty.

Niemcek, for one, said, "It's hard to find a hotel that gives you access to a charging station without asking a real premium for that." Feyen seconds this notion. On a recent trip to St Louis, he found a credit union with a pay charging station (a ChargePoint station, in fact). It was very expensive—$3 per hour—and very slow. "They need to charge by how much energy you take, not how long you're charging," Feyen said. "I'm happy to pay what it costs, but not a premium for using it. I think a lot of people see dollar signs here."

MYTH: EVS ARE NO GOOD FOR ROAD TRIPS

In early 2017, AAA reported that one-third of Americans were planning to travel 50 miles or more from their homes that year.[19] And 79 percent of those travelers were planning to drive. In other words, the great American road trip isn't going away anytime soon.

But is a cross-country journey possible in an EV? Absolutely—although the experience will be a bit different. One of the biggest road trip game-changers: You guessed it, charging stations.

Depending on the route and destination, EV charging stations could be far and few between. Instead of simply stopping at the first gas station on the road to fill the gas tank, hit the restrooms, grab a snack, and get back on the road, EV drivers will need to map out each charging station they'll visit in advance. In addition, drivers will need to consider the number of miles between the stations, and how many miles of range their EV has.

And those range calculation estimations should be conservative.

"One of the first, most valuable lessons in planning is that you should throw out any expectation of getting the same sort of driving range that you've been seeing around town," warned Bengt Halvorson of Green Car Reports. "Electric cars are the most efficient at low speeds, with gentle acceleration. Subject them to the standard Interstate cruising speeds where gasoline engines aren't far off their efficiency sweet spot, and the EV just isn't going to return anything close to its peak range."[20]

Charging times are another sharp contrast between traveling in an EV and road trips in an ICE vehicle, when filling up the tank only takes minutes. And, of course, the speed of each station's chargers will be an important consideration. Charging times range from about 30 minutes to as long as 12 hours, depending on the vehicle, the charger, the state of the battery (drained or partially charged), even the ambient temperature.[21]

Some tourist destinations see EVs' longer charging time as an opportunity to draw visitors: As of this writing, charging stations have already started to appear in tourist-centric locations, from Disneyland Resort in Southern California to the Smokey Mountain Brewery in Tennessee. Atlantic City offers an eco-friendly parking garage and mixed-use center with a rooftop solar array, energy efficient lighting system, and commercial charging stations. EV owners are encouraged to charge their vehicles while shopping, gambling, and sightseeing on the Boardwalk.

Bottom line: Road trips will simply take longer in EVs. Drivers who don't want to build hours of charging time into their trips may prefer to use their ICEs for long-distance drives and use an EV as a second vehicle reserved for local use; others may hold off on purchasing an EV altogether.

ROAD TRIP RESOURCES

Just like ICE vehicle drivers who loaded up on travel planners and roadmaps from AAA a generation ago, EV drivers will have resources at their disposal for road trips.

One of the most critical will be charging station–locating apps. The good news here is multiple apps are available to help drivers locate the best option closest to them. But they're not entirely reliable: App providers sometimes fail to include the competition on their maps.

Two companies, PlugShare and ChargeHub, are working to solve this issue by linking drivers to all known stations via their apps, says Eric Schaal of *FleetCarma*. "There is still some variability in the stations that appear on the maps though, and drivers may have to use multiple—and sometimes warring—apps simply to find an available plug in a convenient location. This situation is making it harder on drivers and undoubtedly slowing the adoption of electric vehicles."[22]

MYTH: EVS DON'T WORK FOR BIG FAMILIES OR LARGE, HEAVY LOADS

Niemcek says he sees few EVs in his neighborhood: "I'm rural, and it seems that everyone around me wants to be in a big car: pickup trucks and SUVs, both of which are becoming more energy-efficient. There are families who need the space in their vehicles, but around here, the status symbol is a four-door pickup truck as the family car."

And for the most part, the revival of the EV over the past decade has brought about compact sedans and hatchbacks, rather than the revered SUVs and pickups.

These vehicles are not renowned for their spaciousness: For one thing, the location of the battery pack (between back seats) cuts into the available seating. The first-generation Chevy Volt, for example, only seats two in the back; newer models added a fifth seat, but it's very small. As one reviewer noted, "To call it a seat would be a stretch; it's little more than a cushioned pad over the battery pack, with the wide battery tunnel requiring the occupant's legs to splay into the foot wells on either side."[23]

"The back seats are cramped. I wouldn't recommend sitting back there to any large people," Feyen said. "If I'd designed it, I would have stretched out the car by six inches."

Niemcek agrees. "Storage space is a problem, and the 'fifth seat' in the back is really only for very small people. My thought about electric cars in general is that it feels like designers run out of energy, and the tail end is an afterthought."

Getting the Word Out

In 2013, *The Wall Street Journal* described EVs as "still such a novelty."[24] By the end of 2018, 1 million EVs had been sold in the United States.[25] And while "the usual suspects" like California and Oregon are still seeing the biggest percentage of ownership, the trend is moving across the country. Even Iowa, smack-dab in the middle of it all, is racking up its share of EV sales that would make inventor William Morrison proud!

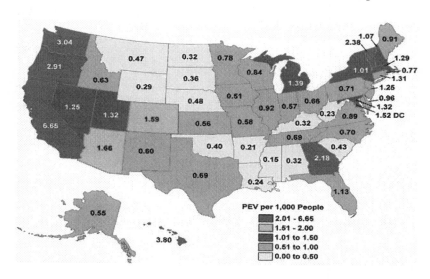

Figure 9.1. *PEV registrations per 1,000 people by state, 2016.*
Source: United States Department of Energy, https://www.greencarreports.com/
news/1114242_these-six-states-have-the-highest-electric-car-adoption-rates-in-the-country.

It's clear that—despite the myths, the points of confusion, the uninformed dealerships, and the inexperienced technicians—EVs are catching on. And one critical component of continued EV success is public education and awareness. After all, consumers who are familiar with EVs, their benefits, and what ownership might be like are more likely to consider buying them.

On the public education front, multiple companies, organizations, and individuals have been excelling in recent years. Electric car clubs, for example, have been giving members a taste of the EV driving experience without the need to commit to a purchase. In the United Kingdom, E-Car Club members have access to a fleet that includes the Renault Zoe, Kangoo Maxi, and the Nissan Leaf. The club offers pay-as-you-go and subscription memberships that include a one-time £50 fee, £15 monthly fees for subscribers, and usage fees ranging from £4.50 per hour or £35 per day to £7.50 per hour or $60 per day.[26]

"We believe E-Car is being launched at a very exciting time for motorists, as many want to experiment with driving an electric car before they make the decision to own one," company chairman Andrew Wordsworth told *Fast Company*. "We hope to grow the E-Car network over the coming months."[27]

Drive Electric Northern Colorado (DENC) is known for helping community members and businesses hold "rides and drives" where visitors can test-drive multiple EVs. "These events have been highly important in influencing regional EV education and adoption, and have encouraged thousands to get behind the wheel of an EV," DENC states on its website.[28]

Taking this approach one step further, several organizations are combining EV-awareness campaigns with tourism.[29] The results include Oregon's Plug and Pinot program and Canada's Okanagan Eco Wine Tour. Oregon's program launched in 2014, when the founding members of the Oregon EV Byway Alliance decided to capitalize on the growing number of wine businesses in Oregon with charging infrastructure. Alliance members teamed up with auto manufacturers to provide EV rides for participants, all with the goal of showing the public that a full-day trip through Oregon's wine country was entirely possible in an EV.[30] During the tour, still going strong today, visitors can test-drive EVs at each destination and talk to tour volunteers about sustainable driving.

There's also the E-mazing Race, created by Sun Country Highway in Canada. During the month-long event, racers drive their electric or hybrid vehicles to Sun Country charging stations across Canada and take photos of themselves there using Sun Country's app. Racers collect points each time they check in, and those with the most points win. Participants are awarded prizes for the best check-in photo and the best social media campaign footprint.[31]

More public awareness standouts include National Drive Electric Week. This U.S. program plugs the benefits of all-electric and hybrid vehicles every September. Events are held throughout the country, including parades, ride and drives, electric tailgate parties, award programs, informational booths, and more. Supporters include Plug In America, the Sierra Club, and the Electric Auto Association.

Like any emerging technology, it will take time before consumers have all the information they need to separate myth from fact and decide if EVs are the right choice. If sales continue, it's likely that American consumers will feel more informed and less apprehensive about making the switch. Learning curves will become less steep. Salespeople and service techs will become more knowledgeable. And perhaps one day the Chevy Volt will get an owner's manual of its very own.

Until then, early adopters like Bradley Niemcek and Joshua Feyen will continue to spread the word and get consumers charged up, as it were, about EVs.

CHAPTER 10

COSTS AND SAVINGS:

HYPE VERSUS REALITY

In his personal account of his 2012 Chevrolet Volt after a year of ownership, Detroit-based environmental journalist Doug Elbinger wrote, "What people really want to know is . . . does it save you money?"[1]

That's a good question, and it's a question that can be hard to answer quickly or easily. Most proponents of EVs will tell you that, in addition to saving the planet and the environment, EVs can also help consumers save money. Investing in EVs, they say, means significant savings, from tax credits and government incentives to drastically reduced costs for fuel and routine maintenance.

But this may be a bit of an oversimplification. While it's true that EV owners will save quite a bit of cash on trips to the gas pump, it's also true that they'll pay more—thousands of dollars more—to purchase their fuel-saving vehicles. EV owners will sidestep some of the routine maintenance associated with ICEs, but they are essentially replacing one set of costs (oil changes; transmission and brake maintenance) with another (replacement batteries; higher repair costs).

So, how do the savings really stack up? Does a "green" vehicle actually equal more "green" in your wallet? Or is this just another example of the hyperbole so often encountered on both sides of the EV versus ICE debate?

Put simply: *Will* an EV save you money?

Let's take a closer look.

Maintenance Costs: Goodbye, Oil Changes . . . Hello Savings?

Those firmly in the pro-EV camp are quick to point out lower maintenance costs as a point of savings. The EV engine sports fewer moving parts than its ICE counterpart. And fewer moving parts means, in theory, fewer places to break down. EV drivers report lower routine maintenance costs because their cars have made things like oil changes, spark plug replacement, and braking system repairs all but obsolete. (If not obsolete, then at least needed much less often.)

EV drivers go a *long time* without needing any of the basic maintenance they were accustomed to getting for their ICE vehicles. One driver even reported the first scheduled maintenance wasn't until 150,000 miles.[2]

But how much can a typical EV driver save when they bypass an oil change or an emissions inspection? The answer is, actually, fairly encouraging: Most studies suggest that maintaining an EV costs about one-third of the cost of maintaining an ICE of similar size.

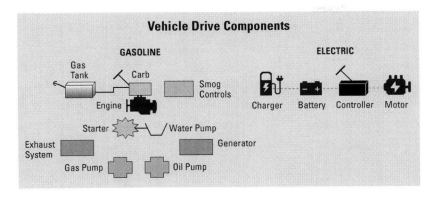

Figure 10.1. *Vehicle drive components, EV versus ICE.*
Source: Idaho National Lab, https://avt.inl.gov/sites/default/files/pdf/fsev/compare.pdf.

An EV's engine is more streamlined, with fewer moving parts than its ICE counterpart. And fewer moving parts means fewer things to monitor and/or replace. ICE drivers have come to expect four regularly scheduled maintenance issues.

OIL CHANGES

Most ICE manufacturers recommend an oil change every 3,000 to 5,000 miles. The average American drives 13,476 miles annually,[3] which means ICE owners can expect three to five oil changes each year at $20 to $55 a pop.

EV owner Bradley Niemcek paints us a picture of just how little oil EVs need. "I bought the car with 1,100 miles on it, and I've put 23,000 miles on it. After fifteen months of ownership, I haven't yet needed an oil change," Niemcek said in a personal interview.

Chevrolet sends an automated alert when the car's onboard computer senses that fluid levels are low, he explained, and so far, the car hasn't indicated the need for a top-off. In fact, Niemcek hasn't needed any service except for getting snow tires put on to handle the Midwestern winter.

SPARK PLUGS AND ASSOCIATED WIRING

When you turn the key in the ignition of an ICE vehicle, the ignition coil sparks and ignites the air-fuel mixture in the engine cylinders to start the engine. These spark plugs wear out over time, and ICE owners should expect to replace them at least once every 100,000 miles. Spark plug replacement can be a labor-intensive proposition, and, depending on make and model, it can cost upward of $200.

TRANSMISSION REPAIRS

Because of the sheer complexity of this engine component, the cost of a transmission replacement or repair can range anywhere from $1,800 to $3,500, depending on the vehicle.[4]

BRAKE SYSTEM MAINTENANCE

Keeping an ICE brake system in solid working order requires regular replacement and maintenance. Brake pads, for example, need to be replaced between 25,000 and 75,000 miles. Rotors and calipers need occasional service, too. All told, average brake system costs look like this:

Component	Parts	Labor	Total
Brake pads	$50–$150	$100	$150–$250
Calipers	$50–$100	$100	$150–$200
Rotors	$200–$400	$150	$350–$550

Estimates from autoservicecosts.com/brake-pad-replacement-cost.

EVS AND "REGENERATIVE BRAKING"

In a traditional ICE braking system, brake pads produce friction with the brake rotors to slow or stop the vehicle. With an EV's regenerative braking, on the other hand, the vehicle's drive system actually does most of the braking: As an EV driver steps lightly on the brake, the onboard computer tells the electric motor to spin more slowly and shift into reverse.[5]

In soft braking, as in stop-and-go city driving, the mechanical brake doesn't engage. EVs do have friction brakes as well, for fast-stopping situations where regenerative braking simply won't supply enough stopping power rapidly enough (i.e., stomping on the brakes to avoid sudden collision). So, yes, EVs sometimes need brake system maintenance, but at much lower instances.

Regenerative braking offers a twofold benefit. First, because the mechanical brakes are not applied very often, there is less wear and tear on the hardware. EV owners will pay less for their brake systems over the lifetime of their cars, with brake pads lasting 100,000 miles or more.[6] Second, applying the brakes actually recharges the battery. While running in reverse to slow the vehicle, the motor becomes an electric generator that produces electricity for the battery. Granted, it's probably not sufficient enough to allow drivers to skip charging, but it's certainly an added benefit.

So, if we are looking at maintenance costs alone, EVs have a competitive edge—with two important considerations.

First, it's worth noting that this advantage might be a bit of a double-edged sword. While less maintenance means more money in EV drivers' wallets, it also means that, when it does come time for a needed repair, EV drivers may struggle to find experienced, knowledgeable service technicians (remember, they're "unproven technology").

Second, it's important to consider battery cost. While the American Chemical Society predicts that well-managed EV batteries may last up to twenty years, it's too early to know if such a prediction will hold true. It's still too early to know what "average" replacement costs are for EV batteries—especially in light of the steadily dropping price of batteries. In mid-2014, Nissan announced that the cost of a replacement lithium-ion battery pack for the Leaf was $5,499 (after factoring in a $1,000 credit for the old battery pack[7]), plus taxes and installation fees for approximately three hours of labor.

But even taking batteries into account, EVs still come out on top.

Stating the Obvious: Fuel Costs

Regardless of the type of EV, one thing that owners universally appreciate is not buying gas (none at all for BEV drivers or much less than PHEV drivers are used to). As noted, PHEVs do require some amount of gas to run the combustion engine after the battery is depleted. But thanks to the battery plus the engine's efficiency, these cars are designed to achieve higher mileage than comparable ICE cars.

And from a series of personal interviews with PHEV owners, it turns out that the EV does just that.

"After 23,000 miles, I have averaged 56.2 mpg-e (electronic equivalent)," said PHEV owner Bradley Niemcek.

"When we run out of electrons, even when we're in gas mode, we still get 40 miles per gallon," PHEV owner Joshua Feyen said. "I also have a 2004 Honda—I love that car, and at the time I bought it, I thought it was great mileage. But now I see that it doesn't come close."

And the PHEV gas tank is much smaller than a regular ICE—meaning a much smaller cost to "fill 'er up." In addition, the rapidly advancing technology seems to be continuing the trend toward spending less at the pump: The first generation of the Chevrolet Volt from 2010, for example, recommended premium gasoline (to promote a cleaner engine), but the second generation in 2015 eliminated that need.

But are the fuel savings alone enough to justify going electric? How much can you really save by making the switch?

Idaho National Laboratory summed it up like this: "Based on an electric vehicle efficiency of 3 miles/kwhr and the cost of electricity at 7 cents per kwhr, the electric vehicle will travel about 43 miles for $1.00. Based on an average of 22 mpg for gasoline vehicles and a gasoline cost of $1.25/gal, the gasoline-powered vehicle will go about 18 miles. Thus, the distance that can be traveled for a fuel cost of $1.00 is more than twice as far with an electric vehicle."[8]

According to *Consumer Reports*'s 2017 Annual Auto Survey, low fuel costs—combined with the lower cost of routine maintenance—could potentially offset the cost of buying a Nissan Leaf in about a year (although the article encourages readers to do their own homework using one of the many EV calculators available online).

Of course, it's important to keep in mind that these savings are estimates. An individual consumer's fuel savings may vary quite a bit, depending on factors such as gas and electricity prices in your area and how many miles you drive in a typical week.

And your satisfaction with your fuel savings may vary, depending on your reasons for choosing to drive an EV in the first place.

What's Your Motivation?

Ask any EV driver why he or she made the decision to go electric, and you'll likely discover that the driver's motivations run the gamut, from assumed savings on gas and "going green" to politics and patriotism to the "cool factor" that comes along with being an early adopter.

"I owned a turbo diesel Volkswagen Jetta—one of those with the completely fabricated 'low-emissions' models that the company lied about," Niemcek said. "After several years driving that ecologically devastating thing around, I wanted to do something better, something with a lower impact on the environment. It just seemed like the cool thing to do."

Interestingly, Volkswagen was part of Feyen's decision-making as well. "I had the VW diesel Golf that was part of their naughty polluting diesel cars. I felt like I had some sins to repay after driving around in that horrible car for four years," he explained. "I had bought that one deliberately—I'd studied cars for a year and a half before buying it. It got good mileage, and it was a fun car to drive. So I wanted to replace it with something that was equally good—or even better—environmentally."

Feyen, an urban farmer and community builder whose "day job" is in social media with American Family Insurance, also put solar panels on his house almost concurrently with buying his EV.

"The idea was, if I'm buying an electric car, what can I do to fuel that car? I can make my own electricity; I can't make my own gas," he said. "Now I know where the fuel comes from, and I know how much it costs—I feel good about it."

For Mark Clayton, a writer for *The Christian Science Monitor*, the "feel-good" part of driving an electric vehicle was less about pollution and more about energy independence. Because EVs run on domestically sourced energy, their drivers are limiting the use of imported oil and leaving more U.S. oil in the till to be exported.

In a series of articles about his experience as the owner of a Nissan Leaf, Clayton wrote that his decision to go electric was a personal, patriotic one: By driving an EV, he and his wife were doing their part to reduce U.S. dependence on foreign oil. "The real reason Laura and I took possession of an 'ocean blue' Nissan Leaf on an icy day in mid-February was 9/11. After that terrible day, many people began asking: What can I do in my personal life to make sure this never happens again? Not a lot, really, I thought at the time," Clayton remembered. "But lately my view has changed. One thing is that it might help to refuse to give 'people who hate us' money for their oil."[9]

Although he's quick to admit that this line of thinking is a bit "simplistic," Clayton said an all-EV future could mean an energy-independent future where the United States can rely 100 percent on domestic fuel.

Clayton and his wife don't drive much outside of a 50-mile radius, and they save about $1,000 per year on fuel. That's not bad, but he says that his monthly lease payments basically cancel out the savings.

But for Clayton, the principle of driving an EV is more important than his out-of-pocket costs at the pump.

Sale Price and Resale Value: Spending (and Losing) Money to Save Money?

EVs cost more money to buy than ICEs. A lot more. While nowhere near exhaustive, this 2019 pricing guide from EV Rater (Figure 9.2) provides a snapshot of current MSRP (plus any mandatory destination and handling fees) for EV options currently available in the U.S. market.

MAKE / Model	Price	Range / miles	Battery / kWh
AUDI e-tron	$74,800	204	95
BMW i3	$44,450	114	33
CHEVROLET Bolt	$36,620	238	60
FIAT500 e	$32,995	84	24
FORD Focus Electric	$29,120	115	33.5
HONDA Clarity Electric	$37,510	89	25.5
HYUNDAI Ioniq Electric	$29,500	124	28
HYUNDAI Kona Electric	$36,450	258	64
JAGUAR I-Pace	$76,500	234	90
KIA Niro	$39,000	239	64
KIA Soul EV	$33,950	111	30
NISSAN Leaf	$29,990	151	40
SMART ED	$23,800	100	17.6
TESLA Model 3 Standard	$35,000	220	55
TESLA Model 3 Long Range	$43,000	325	74
TESLA Model S Standard	$79,000	270	75
TESLA Model S Long Range	$83,000	335	100
TESLA Model X Long Range	$88,000	295	100
TESLA Model X Performance	$104,000	289	100
VOLKSWAGEN e-Golf	$30,495	119	35.8

Figure 10.2. 2019 EV MSRP pricing guide.
Source: https://evrater.com/evs.

While this list does include a wide range of prices, it's worth pointing out that even the least expensive EV may be just out of reach for many Americans: Consider the $29,000-plus Ford Focus Electric and its ICE counterpart, the $18,000 Ford Focus.

That's a pretty big difference.

If you consider MSRP alone, EVs are the more expensive option by far. But that's not the only area where ICEs are in the lead: When it comes to resale value, EVs are at a definite disadvantage.

The National Automobile Dealers Association's Used Car Guide indicated that, in 2014, EV depreciation rates were nearly twice that of comparable ICE vehicles.[10] In fact, a 2019 study by iSeeCars.com compiled their data as an online search engine for used cars and discovered that EVs "won" the top two spots on a list of the highest-depreciating vehicles: After five years of ownership, the Nissan LEAF loses 72.7 percent of its value and the Chevrolet Volt loses 71.2 percent.[11] As EV technology improves and battery range extends, it may be that these "early" models are going the way of the dinosaur. And—we'll talk about tax incentives in more detail in a moment—many tax incentives are only applicable to new cars, which may further discourage buyers from purchasing used EVs.

While resale value might not be top of mind to consumers shopping for a new car, a vehicle's trade-in value can affect the bottom line. For drivers who go into the deal hoping to get something at the end of their car's usefulness, EVs might not yield much payout. Low resale prices are not only bad for current EV owners but also possibly for the EV industry as a whole. If the trend continues, it could push potential new owners away.

Bottom line? If we look only at sale price and resale value, ICEs are the clear winner. But that, too, is a bit of an oversimplification. In some parts of the country, tax credits and other government-sponsored incentive programs are helping to level the playing field.

Tax Credits and Incentives

Here is how the Department of Energy's Alternative Fuels Data Center explains the current EV tax credit:

QUALIFIED PLUG-IN ELECTRIC VEHICLE (PEV) TAX CREDIT

A tax credit is available for the purchase of a new qualified PEV that draws propulsion using a traction battery that has at least five kilo-watt-hours (kWh) of capacity, uses an external source of energy to recharge the battery, has a gross vehicle weight rating of up to 14,000 pounds, and meets specified emission standards. The minimum credit amount is $2,500, and the credit may be up to $7,500, based on each vehicle's traction battery capacity and the gross vehicle weight rating. The credit will begin to be phased out for each manufacturer in the second quarter following the calendar quarter in which a minimum of 200,000 qualified PEVs have been sold by that manufacturer for use in the United States. This tax credit applies to vehicles acquired after December 31, 2009.

This sounds like a pretty sweet deal, especially if you live in a state or city that offers additional perks for going electric. Combined with additional credits at the state level, an EV buyer could potentially snap up savings of close to $10,000—easily closing the price gap between the Ford Focus and the Ford Focus Electric.

At this point, we'll ignore the larger question of whether government incentives are ever a good idea. Whether you agree or disagree with it in principle, there are some very real problems with the language of the current federal EV tax credit.

First, the amount of the tax credit is hardly exact. Note that the description says the credit is for "$2,500 to $7,500"—and the amount of the credit depends on the size and type of the vehicle, as well as the vehicle's battery capacity. Of course, $2,500 is still a significant amount—but it's a far cry from $7,500.

Then, there's the question of what counts as a "qualified EV." The credit is available for the purchase of new EVs only. This means that used EVs—even "certified pre-owned" ones—are not considered "qualified."

But that's not the only point of confusion: All battery-electric vehicles (BEVs) are eligible for the full amount, but not all plug-in hybrids are. For example, consumers who purchased the 2017 Chevrolet Volt received the full credit, but those who bought a 2016 Hyundai Sonata Plug-in Hybrid only got $4,919. To complicate things further, some EVs aren't eligible for *any* tax credit whatsoever. As it turns out, the definition of "qualified" only includes vehicles that can be charged from an external source, regardless of battery capacity.

Of course, none of this may matter at all if current EV sales trends continue. Note this sentence in the description: The credit will begin to be phased out for each manufacturer in the second quarter following the calendar quarter in which a minimum of 200,000 qualified PEVs have been sold by that manufacturer for use in the United States.

Tesla was the first company to hit 200,000 EV sales in 2018. As of July 2019, the amount in federal credits for Teslas was down to $1,875, and on the last day of 2019, the subsidy ran out completely. GM's Chevy Bolt hit 200,000 EV sales right after Tesla. By April 2020, any electric GM sales won't have any subsidy, just like Tesla. The all-electric Nissan Leaf still has the full tax credit.[12]

In other words, the days of federal EV tax breaks may be numbered. But even if tax cuts begin to phase out at the federal level, consumers may be able to take advantage of purchase incentives at the state or municipal level, such as—

- The installation of an EV charger at home (this is on top of the 30 percent of installation cost some homeowners can claim on their federal taxes)
- Additional local tax credits
- Lower charging rates or time-of-use incentives for charging
- More attractive loan options and/or lease rates
- Vehicle or infrastructure rebates or vouchers
- Reduced vehicle registration fees
- Exemption from emissions inspections
- Access to "high-occupancy vehicle" lanes or toll lanes

Every state is different—from the types of incentives they offer to the amount extended. Colorado, for example, offers upward of $5,000 on top of the $7,500 federal credit (if all eligibility requirements are fulfilled, of course).

So, do EV tax incentives live up to the hype? If your EV purchase qualifies and you live in a state like Colorado, the tax credits and financial incentives you receive can bring the cost of an EV in line with its ICE counterpart. But it's important to remember that not every consumer will share this experience.

Insurance Rates: The (Potentially) High Cost of Being an Early Adopter

The information about insurance rates for EVs is, perhaps unsurprisingly, conflicting. Some insurance companies offer lower rates to EV drivers, based on the view that the types of people who purchase EVs are, as a group, "more responsible" than drivers of traditional ICEs.[13]

But this is definitely not a hard-and-fast rule. In fact, one 2015 study compared the insurance premiums for gas and electric versions of the same cars and found that, in all cases, the cost of insurance for the EV was more than the ICE version. Granted, this study was limited in scope: It included rates from the nine largest insurers in one state (California), and it considered only four models with both electric and gas versions.

But in all four comparisons, insurance rates for gas-powered vehicles came out cheaper by an average of 21 percent.

2015 vehicles	Gas	Electric	Difference
Chevrolet Spark	$1,669	$1,982	19%
Fiat 500	$1,597	$2,016	26%
Smart Fortwo	$1,474	$1,807	23%
Volkswagen Golf	$1,741	$2,014	16%
Average	$1,620	$1,955	21%

Figure 10.3. Annual car insurance quotes for ICE and EV in California.
Source: https://www.nerdwallet.com/blog/insurance/car-insurance-quotes-electric-cars/.

The main reason for the disparity? To many insurers, EVs are still considered "unproven technology," and they typically have higher values and higher repair costs. They're also harder to work on, due to the learning curve, which means that insurers may have to seek out specialized repair shops or technicians trained to work on EVs.

Another factor? According to a 2009 study by the National Highway Traffic Safety Administration, EVs have higher rates of accidents involving inattentive pedestrians and bicyclists (a somewhat disturbing downside to the cars' quieter engines).[14]

So, for those keeping score, ICEs have a slight edge over EVs in the insurance department.

What's the Bottom Line?

So, will EVs save you money?

Remember Doug Elbinger and his Chevy Volt? After one year of owning an EV, Elbinger answered the question as follows: "Yes and no, but mostly no. You can purchase a car for equal power, performance, and style for much less money."

Elbinger determined that he was saving about $40 per month in gas (the Volt is a dual battery-gas system) and he usually only filled up once a month. He was impressed even though his mileage results varied greatly from the published Chevy Volt data. But he also cited data from Edmunds.com that "the price premium paid for the Volt, after discounting the $7,500 U.S. federal tax credit, takes a long time for consumers to recover in fuel savings, often longer than the normal ownership time period."[15]

On paper, it appears that EVs offer some very real savings over ICEs. Even with a slightly higher purchase price, EVs offer significantly lower maintenance and fuel costs, making them a more economical choice over the long haul. But it seems that there are still some bugs to work out before that translates to real-world savings.

That day may come sooner than we think. As EV sales continue, they will slowly begin to edge out of "unproven technology" territory.

It's likely that batteries will get cheaper and hold charges longer. As dealerships and service technicians become more familiar with EVs, they'll become less tricky to diagnose and repair. Insurance rates for EVs may fall. Bloomberg New Energy Finance analysts predict that, by 2022, total cost of ownership will be lower for EVs than for ICEs.[16] That lifetime figure combines original purchase price, operating and maintenance costs, and—of course—batteries.

Until then, though, financial rewards alone may not be enough to bring about widespread mainstream acceptance of EVs.

PART III

THE DETAILS

CHARGING, BATTERIES, ROADS, AND . . . PLASTICS?

CHAPTER 11

INFRASTRUCTURE:

IF YOU BUILD IT, THEY WILL COME

Trying to explore all the topics related to EVs is like battling the nine heads of the Hydra. The moment you think you've conquered one, two sprout up in its place. EV-related subjects cascade into a kaleidoscope of discussions, from road maintenance (if EV owners buy no gas, they aren't paying the associated highway taxes) to rare earth mining in Mongolia (batteries require lithium, the extraction of which raises both humanitarian and environmental concerns). The relationship between oil and the lightweight plastics preferred for EVs is also of note, as is the discussion of infrastructure: How will such an enormous network of charging stations be built? More important, who will pay?

Regarding the latter, it's not unusual to hear the argument that the government must fund charging infrastructure in order to make it viable for users to purchase EVs. That's not quite the case, or so we've learned from the forward-thinking Kansas City Power and Light (KCP&L).

Yes, you heard correctly. *Forward-thinking. Utility company. Kansas.*

Technological innovation is probably not the first thing most people associate with Kansas. The rectangular-shaped state has a rather square image, better known for wheat, sunflowers, and NCAA basketball championships than for leading-edge anything. Yet here, smack-dab in the center of the United States, an EV revolution is taking place. And a utility company is leading the charge.

In 2015, KCP&L put down $20 million to install 1,000 charging stations throughout its service territory of more than 800,000 customers.[1] This was despite the fact that, at the time, there were only about 1,600 plug-in electric vehicles in the entire state, a quarter of them in the immediate Kansas City area.[2]

Less than two years later, the utility had not only met its goal; it had exceeded it as well. Today, there are more than 1,000 KCP&L-branded charging stations up and running in places like grocery stores, apartment complexes, and malls. In classic "if you build it, they will come" style, availability of charging infrastructure rapidly spurred EV uptake. The area straddling the Kansas-Missouri border is now one of the fastest-growing EV markets in the nation.

With one of the biggest barriers to EV acceptance out of the way, consumers came rushing in. And it all took place without a government mandate or dinging an unwilling taxpayer.

It's not that KCP&L didn't consult the government or attempt to pass along some of the cost to the public. It's just that when the utility asked the Kansas state legislature for a boost in the form of a two- to three-cent monthly fee added to all customers' bills, the response was a flat-out "no." On the other side of the state line, Missouri legislators also nixed the request.

The utility hadn't proved there was sufficient consumer demand for an extensive network of charging stations, legislators said. David Nickel, consumer counsel with the Kansas Citizens' Utility Ratepayer Board—which represents the interests of utility customers—told NPR's *All Tech Considered* that while KCP&L was free to roll out their Clean Car Network initiative, they couldn't do it by imposing costs on a "captive consumer."[3]

"Let the private sector invest in the EV market, rather than have ratepayers finance the speculative venture," the Kansas Corporation Commission ruled.[4]

That didn't stop KCP&L. They simply footed the entire bill themselves.

Obviously, their interest wasn't in peddling cars—there's no secret Tesla collusion case about to make headlines—but to sell more of what they make: electricity.

According to KCP&L's Chuck Caisley, the utility's nuclear- and wind-driven power grid is underutilized most of the time.[5] Customers have already paid for the grid and the power plants. Getting more people to use more electricity improves efficiency of the infrastructure, which drives down KCP&L's per-unit cost. And that means lower bills for *all* consumers.

The result is a win all around. The Kansas utility removed roadblocks, consumers kept their right to choose, and no one had to submit to government edicts. Best of all, in Kansas, EV drivers are actually helping lower costs across the entire grid while doing their carbon-reducing part.

This story signals how well positioned utilities are to help quickly transform the EV market, a fact that even environmental watchdogs, notoriously tough on utilities, are taking notice of.

As Max Baunhefner of the Natural Resources Defense Council told Stateline writer Martha T. Moore, "Utilities may not be the most innovative companies in the world but they are good at deploying boring electrical infrastructure that doesn't break. That's what EV drivers want and it's sorely lacking at this point."[6]

Utilities: Balancing the Grid

Although KCP&L was an early success story, it's not the only one. In the United States, getting into the charging business has become a coast-to-coast strategy for power companies.

While state approval for their ideas is a must, they're not relying on government funding to get their plans moving or passing costs along to unassuming consumers. Instead, for the most part, they are setting up pay-as-you-get-ready-to-go situations. Charging station users are the only ones who get charged.

The California Public Utilities Commission gave the nod to proposals by three of the state's largest utilities—Pacific Gas and Electric, Southern California Edison, and San Diego Gas & Electric (SDG&E)—to build more than 12,500 public charging stations for about $200 million.[7] SDG&E's Power Your Drive program lets users add the cost of charging their EVs on their electricity bill, and provides time-of-use discounts during what they call grid-friendly hours. The deal is even more attractive for EV drivers who live in what the utility considers a low-income area. For them, the cost of a charge is free.[8]

In Kentucky, Louisville Gas and Electric and Kentucky Utilities have gotten the green light to build as many as twenty charging stations, the cost to be recouped by collecting $3 an hour from the motorists who use them. The state's public utilities commission also said the two companies can install as many charging stations as they want on commercial properties, where property owners will foot the bill.

New Jersey got into the act, too, thanks to a pilot program by Public Service Electric & Gas that installs charging stations at customer locations around the state. The utility provides the charging equipment, and participants pay for installation and electricity.[9]

In Ohio, electric utility AEP partnered with other market participants to accelerate transportation electrification. Although it passed on the chance to put up its own charging stations, AEP is still focused on getting them installed—just by someone else. AEP has asked the state to approve the nearly $10 million package of rebates it wants to give other companies to build chargers, believing that will get the job done faster. The largest rebates—100 percent of the cost—will go for charging stations installed in low-income neighborhoods.

The fact that AEP would offer rebates to companies for building charging infrastructure may sound odd, but it's the natural evolution of marketing by power companies. Utilities have a long history of offering rebates for products that use electricity, including large appliances like washers, dryers, and refrigerators. Why should the apparatus to power a plug-in car be any different?

Or for that matter, the car itself?

In 2017, the Burlington Electric Department, a municipal utility in Vermont, started offering rebates of either $1,200 or $600 on the purchase or lease of a new EV or hybrid. The funding comes out of the utility's cash reserves.[10]

A report by the National Conference of State Legislatures describes what may be the most sweeping rebate initiative by electric utility companies to date. Utilities in Colorado, Delaware, Florida, Georgia, Indiana, Kentucky, Maryland, and Pennsylvania—as part of a special group buy—offered a $10,000 up-front rebate to their customers and employees for the purchase of a new 2017 Nissan Leaf at participating dealerships.[11] It is unclear how much of that came from the utilities themselves, the automaker, or third parties such as energy efficiency or transportation groups.

In addition to rebates, utilities are using preferred rate strategies to get motorists to go electric. Underutilization is a real problem for electricity providers and so is excessive demand during peak periods. Trying to balance the grid is one reason utilities offer time-of-use promotions such as "free nights and weekends," and it stands to reason that consumers would be just as happy to save money charging their cars as they are watching TV or running the dishwasher.

According to Mike Salisbury and Will Toor of the Southwest Energy Efficiency Project, or SWEEP, off-peak charging serves multiple purposes: It avoids increases in peak demand while giving utilities a way to boost demand during those hours they have a large amount of underutilized capacity.[12]

Getting Africa on Board

While utilities are helping speed up EV adoption in the United States, the kind of solutions they offer are workable only where there is a strong, intact grid. For the most part, Africa is not one of those places.

Yet, here, too, progress is being made without the involvement of bureaucrats. Instead, international investors are giving African countries and cities a hand up—and a step toward a cleaner future.

For example, the large French investment company Bolloré introduced Cameroon's first electric bus line in 2014. The buses, which run
on solar panels connected to electric batteries, ferry faculty, staff, and
students around the University of Yaoundé.[13] They've been such a hit
that Cape Town, South Africa, after taking bids from multiple international electric bus manufacturers, chose China's green energy firm BYD
to deliver 11 vehicles by the end of 2017. BYD, which is backed by American billionaire investor Warren Buffett, signed an agreement to set up
shop in Morocco, with plans to open a factory for electric cars, buses, and
trucks by the end of 2018. The Chinese firm also has a facility in Hun-
gary and has invested €10 million to build a plant in France.[14]

Ten of the Cape Town buses will circulate in the city; the eleventh
will be on loan to the Namibian city of Windhoek, which is mulling the
acquisition of its own fleet. To offset the coal-generated electricity consumed by the new buses, the City of Cape Town plans to install solar
power at bus and maintenance depots and bus stations.

Investors Give Clean Energy Firms a Jolt

Investing in EV technology isn't strictly the province of companies like
Bolloré or rich financiers like Buffett. With the emergence of clean
energy banks, investors of all stripes are getting a chance to participate
in financing energy efficiency and renewable energy efforts.

One way is through clean energy banks (CEBs), which help green
energy firms access lower-cost capital from the private sector, with no
risk to taxpayers. CEBs take the place of larger, more traditional financial
institutions that aren't willing to invest in clean energy opportunities,
often because they aren't familiar with the technologies or the deals are
too small for their portfolios.[15] CEBs use limited public funds to attract
and educate investors about potential investments and standardizes
financial products to make them simple to buy and sell. The investors—
including corporate entities, pension funds, endowments, foundations,
and social impact stakeholders—provide financing directly to the clean
energy firm.

No wonder CEBs have been described as "matchmakers helping commercial investors fall for clean energy projects by revealing their unseen potential."[16]

In the United States, the granddaddy of state-legislated CEBs is Connecticut Green Bank, launched in 2011. Connecticut Green Bank says that its public-private partnership model has "upended the government subsidy approach to clean energy."[17] In less than seven years, the bank and its private investment partners had deployed more than $1 billion in capital for clean energy projects across Connecticut.

According to the Green Bank Network's website, the CEB model is growing in the United States—as of December 2018, 13 other banks have followed Connecticut Green Bank's lead, including in California, Hawaii, Maryland, New York, and Rhode Island.

For the analysts at C2ES, The Center for Climate and Energy Solutions, the CEB modus operandi of leveraging public funds to attract private investment in charging infrastructure is a no-brainer.

Through a combination of low-interest or longer-term loans, interest rate buydowns, project equity stakes, small grants, and, as the market develops, credit enhancements, C2ES says that CEBs are able to increase overall investment and accelerate technology deployments.

"By drawing on their experience in building efficiency and other clean energy projects, clean energy banks can be part of the solution," C2ES wrote in a 2014 report.[18]

Although the exact CEB format hasn't been duplicated in other countries, different forms of clean energy financing have emerged all over the world:

- Energy Impact Partners (EIP), the bicoastal American company with offices in New York and San Francisco, made its first-ever e-mobility investment by backing Greenlots, a leading provider of grid-enabled EV smart charging solutions. EIP's confidence in Greenlots has already paid off: Early in 2018, Greenlots was selected to provide the operating platform for Electrify America's network of high-power fast chargers—part of a $2 billion investment in EV infrastructure. Greenlots's operating platform will help drivers instantly locate the closest charger, know their charging status, and quickly make payments.[19]

• The European Investment Bank (EIB), the world's largest multilateral borrower and lender, in 2017 approved new financing for €4.3 billion for renewable energy and security of energy supply projects. EIB says it commits at least 25 percent of its lending portfolio to low-carbon and climate-resilient growth around the world. In 2016, that came to €16.9 billion.[20]

• In Japan, two of the nation's largest lenders, Mitsubishi UFJ Financial Group and Mizuho Financial Group, provided more than $3.2 billion in loans for clean energy projects in the first half of 2017 alone.

• Since 2014, when 40 banking and development experts helped create China's Green Finance Task Force, the government has been issuing green bonds as part of their goal of generating $600 billion in private sector financing.[21]

A Surge in Chargers Built by Automakers

It's not clear now how many private investors will be lending the money to build charging stations, but making sure that there's always somewhere to "juice up your ride" is going to be big business, says *MIT Technology Review*. The publication cites estimates that the world will need to spend $2.7 trillion—that's *trillion*, with a "T"—on charging infrastructure if it's to support what might eventually be as many as 500 million EVs.[22]

With lack of charging density making consumers squeamish about buying an EV, it makes sense that EV automakers would try to allay fears by building their own public infrastructure. Tesla's off to a significant head start in that regard: As of May 2019, it has 1,441 Supercharger Stations with 12,888 Superchargers worldwide, which it runs and manages without outsourcing.[23] (To put these figures into perspective, these figures have grown since January 2018, when there were 1,130 Supercharger Stations with 8,496 Superchargers.)

Tesla's Supercharger fast-charging network was originally designed to enable longer-distance travel, but the company is working to accommodate the younger buyers who make up the new car's intended market—drivers who might not have charger access at work or at home. For them, the company is installing more of its Superchargers in cities. By the end of 2017, Tesla had 10,000 fast-charging Superchargers and 15,000 of its slower Destination Charging connectors globally. The number of Tesla charging locations in North America alone was supposed to grow by 150 percent in 2017.[24]

Nissan and BMW are also building infrastructure; in fact, they were setting up chargers for urban commuters when it was still just a spark of an idea for everyone else. But now, in a surprising twist, the two automakers have turned their attention to day trippers and other longer-haul drivers.

In 2017, Nissan added nine more stations along the I-95 corridor between Boston and Washington, D.C., in partnership with charging provider EVgo. The nine charging sites—which opened in time for the launch of Nissan's all-new Leaf EV—include 50 total chargers that can replenish four or more EVs simultaneously, in slightly more than 30 minutes each.[25]

Nissan also partnered with BMW to build 174 DC fast-charging stations across 33 states in 2016, plus another 50 in 2017.[26] That brought the total network to more than 668 stations in more than 50 metro areas.[27]

"You can't just make the car and walk away from the infrastructure," Robert Healey, head of EV infrastructure for BMW, told the *New York Daily News*. "We support an infrastructure build up, which we've been involved with since 2008."[28]

In 2017, BMW joined forces with the National Park Service, National Park Foundation, and the Department of Energy to install charging stations at what might be the most appropriate venue of all time: the Thomas Edison National Historic Park in West Orange, New Jersey. The site's four level 2 chargers are the first of as many as one hundred EV charging stations planned for in national parks and nearby communities.[29] It takes roughly two to three hours to power up an EV at Thomas Edison Park. Coincidentally, that's about how long it takes to tour the Edison Museum.

BMW is also working to make an infrastructure impact in Europe as well. Its collaboration with Ford, Daimler, and Volkswagen—called Ionity—plans to install a network of 400 fast-charging stations across the continent by 2020. As of May 2020, 226 stations are live and 51 are under construction.[30] The move appears to be a direct assault against Tesla, which has 350 European Supercharger stations.

For Volkswagen, the Ionity alliance isn't its only foray into the charging game. In a sense, though, it took some strong-arming to get them to suit up: The company committed $2 billion to developing charging points in a settlement over its fraudulent diesel emissions scheme.[31] Volkswagen's environmental mitigation trust is also helping fund other EV incentives. Colorado, for example, will use some of the $68.7 million it receives from the Volkswagen trust to lure motorists into purchasing EVs and has earmarked part of the windfall to develop ZEV fueling and charging infrastructure.[32]

Over the EV Rainbow

Back in Kansas, KCP&L's Clean Charge Network chargers are a popular fixture around town, not surprising given that the metro area experienced 78 percent growth in EV adoption between 2016 and 2017. That was more than any other city, including green hot spots such as Los Angeles, Denver, and Durham. A dedicated Clean Charge Network website not only informs visitors where to find a charger, but it also leads them to the perfect EV, thanks to a BuzzFeed-y quiz with serious questions ("How much stuff do you have to haul each day?") and whimsical responses ("I have lots of baggage, literally and metaphorically.").

If the hipster swagger of answers like that doesn't quite jibe with the lasting image of Kansas, so be it. The future is here, right here in Kansas City, and the town is embracing it. And, to trot out another Kansas trope, it didn't take an omniscient, Oz-like government wizard to make it happen. Pull back the curtain of subsidies and schemes, let the open market run the show, and a gray and cloudy EV picture opens up in living Technicolor.

CALIFORNIA

The Los Angeles Department of Water & Power's "Charge Up L.A.!" program offers residents a rebate of up to $500 toward charging equipment and installation.

Burbank Water and Power offers rebates of up to $500 for residential charging station equipment and up to $1,000 for commercial stations. In addition, the utility discounts the per kilowatt-hour rate for use of public EV charging locations during both peak and off-peak hours.

In June 2009, Sony Pictures Entertainment established an employee incentive of up to $5,000 when they buy either a hybrid electric vehicle or install solar voltaic panels on their residence.

GEORGIA

Georgia Power offers three different charging rate options for residential customers. The Plug-In Electric Vehicle rate provides a discount on electricity when used from 11 p.m. to 7 a.m.

MICHIGAN

The Lansing Board of Water & Light offers an experimental residential EV charging rate. Indiana Michigan Power offers a specific EV charging rate. Consumers Energy offers a time-of-use EV charging rate, with reduced prices between 11 p.m. and 7 a.m. DTE Energy also offers two specific EV charging rates.

PENNSYLVANIA

Clearview Energy offers a $75 rebate toward a new ChargePoint residential EV charger with enrollment in Clear Charge Energy Plan.

SOUTH CAROLINA

Duke Energy offers funding of up to $5,000 per charging port, $20,000 per site, or $50,000 per city for EV charging.

WASHINGTON

Puget Sound Energy offers a $500 rebate to customers who purchase and install qualified level 2 chargers.

CHAPTER 12

CONFLICT CHEMISTRY:

RARE EARTH MINING AND RANGE ANXIETY

It is entirely possible to charge an EV battery in five minutes.

"This is happening; this is real," said Doron Myersdorf, CEO of Israeli startup StoreDot, during a 2016 talk about his business's battery capabilities.[1] StoreDot's FlashBattery uses layers of nanomaterials and proprietary organic compounds, Myersdorf explained. "We are designing a new generation of batteries, essentially rebuilding across all four components that currently make up lithium-ion batteries: the anode, the cathode, the charge transfer and the separator put in place to prevent shorts," he told TechCrunch in 2017.[2]

The availability of a five-minute charge would be a big deal. Huge. Currently, the fastest charge available—from a fast-charging, level 3, public station—takes about 30 to 40 minutes. And that's assuming the driver finds a public spot with fast charging—and it works with their particular car. The CHAdeMO network only fits Japanese and Asian-made vehicles. The SAE Combo plugs only fit German and American cars.

Tesla's Supercharger network is only for that company's vehicles. Drivers who find level 2 stations may spend three to eight hours charging an empty battery.[3]

Not only can StoreDot's new battery reach full charge within five minutes. The company also claims the charge lasts for 300 miles. Myersdorf notes that, normally, charging a battery that quickly can be dangerous. The graphite used in lithium-ion batteries overheats when charged too fast and can even cause fire or an explosion. But StoreDot's technology can hold a charge without heating up.

The five-minute charging technology has yet to go to market. The company has, however, raised $60 million in funding from the likes of Daimler and Samsung, and it has been ranked among the one hundred most disruptive startups in the world.[4] StoreDot is now in the process of obtaining regulatory approval for its battery. And at the end of 2018, StoreDot entered into a partnership with Chinese EVE Energy Co. to begin mass production of the FlashBattery.[5] The website lists BP and TDK Corporation as the company's other partners.

StoreDot's technology is just one of numerous technological innovations and creative programs in recent years that genuinely have been making charging easier, faster, or more convenient, bringing the world closer to greater EV acceptance and usage.

Although not the only battery option, lithium-ion batteries are the battery of choice for EV manufacturers. With a density of 1,700 watt-hour per kilogram, these batteries could power an EV for roughly 311 miles on a single charge.[6]

They are currently superior to other technologies. Roughly speaking, these batteries deliver about four times the energy density (the amount of energy stored in a given area, such as a battery) of lead-acid batteries commonly used for gasoline-powered cars. Lithium-ion also produces roughly twice the energy density of the nickel–metal hydride (NiMH) batteries originally used in the first generation of electric cars.

The lithium-ion industry is already a powerhouse: The market is expected to experience significant growth in the next few years, reaching $40 billion by the year 2025.[7]

But how did the lithium-ion revolution begin? To answer that, we'll have to go back to nineteenth-century France, where a physicist named Gaston Plante invented the first lead-sulfuric acid battery in 1859. Other scientists began improving on his design almost immediately, and a technique for manufacturing the batteries came within a decade. As early as 1902, Thomas Edison announced the development of the iron-nickel battery and, a few years later, optimized its performance by adding a lithium hydroxide compound to the electrolyte (a gel, liquid, or solid material enabling electricity to flow between the battery's positive and negative electrodes).

Decades of intense experimentation followed to increase power, and by 1972, Exxon Enterprises hired a British chemist, M. Stanley Whittingham, who developed a rechargeable titanium-lithium battery with the potential of storing significant amounts of energy—about 480 watt-hours per kilogram. However, the battery materials were too expensive, and lithium-metal combinations were prone to explode or catch fire. By 1978 and 1979, however, researchers at Stanford and Oxford University demonstrated a rechargeable lithium cell with a 4-volt range using lithium cobalt oxide ($LiCoO_2$) in the positive electrode (the cathode) and lithium metal in the negative electrode (anode).

The discovery of the stable $LiCoO_2$ compound opened up the potential for novel rechargeable batteries, and several important modifications and improvements followed.

Today, lithium-ion batteries are valued not only for energy and power density, but also for the diversity of hardware and electronic components they safely power. Lithium-ion energizes cell phones and laptops, electronic readers and iPads, game systems, power tools, and, of course, hybrid/full electric vehicles. Various electric grid applications for energy storage can also employ large stacks of lithium-ion batteries to improve the usability and amount of energy harvested from wind, solar, geothermal, and other sources.

In short, lithium-ion is probably the most fungible tool we have today to build a sustainable energy economy.

That said, all is not perfect in paradise.

No Myth: Common Misconceptions About Lithium-Ion Batteries

There are many pervasive mythologies—urban legends and tall tales—associated with lithium-ion batteries that are simply not borne out by facts. At least one of them is highly controversial: How much improvement in battery packs can we expect using current lithium technologies now and in the near future?

The pervasive wisdom is that improvements in energy density are predictably unlimited.

EV pioneer Tesla, for example, claims that battery energy density doubled in the decade between 1995 and 2005, but this might be a bit misleading: While it's true that Tesla has steadily increased the density of its individual lithium-ion battery cells, *improvements to the entire battery pack are much more modest.*

Over the past few years, the maximum pack level energy density for a Tesla lithium-ion battery has hovered at a relatively stable 150–170 watt-hour per kilogram (although some critics argue that "stable" is another way to say "flatlining," or not improving at all). Still, the batteries for Tesla's smaller Model 3 "economy car" have been engineered to drive more than 210–315 miles on a single charge. A 2017 pure electric Chevy Bolt, by comparison, claims a range of 238 miles on a single charge.

Both ranges are impressive, and they have been increasing steadily (although the EPA *Certification Report* of July 2017 does not bear out Tesla's claim). However, there is reason to believe that lithium-ion chemistries have their limits, particularly considering their comparatively slow energy density growth in the past few years. And as attorney and energy author John Petersen noted in a critique of Tesla's calculations, "Discussions of cell level energy density . . . are interesting to battery geeks but meaningless in the real world because a lithium-ion cell can't do any useful work until it's built into a complete battery pack that can power a vehicle or store electricity from a solar panel."[8]

A new technology of batteries such as lithium air or aluminum air, along with other energy sources for EVs—perhaps hydrogen fuel cell, if costs of platinum catalysts can be reduced—may be necessary to accelerate range to much higher levels (more on new battery chemistries later).

Of course, this is just one misconception about lithium-ion batteries. Let's examine a few of the more widespread ones.

Myth: Choosing an EV is just trading off the same amount of greenhouse gas pollution from tailpipe to dirty electric utility.[9]

REALITY

As we've discussed, a "wells to wheels" analysis shows that an electric car will produce significantly less carbon dioxide pollution from electrical charging than the CO_2 pollution from a conventional gasoline-powered car. The carbon imprint will decrease with long-term driving. If the source of the electricity for recharging is from a utility relying partially or wholly on renewable sources, or even natural gas, the emissions benefit is even greater.

Myth: Lithium-ion batteries pose a recycling problem.

REALITY

It is true that smaller lithium-ion batteries often find their way to the dump. In Europe, only about 5 percent of lithium-ion batteries from small electronics such as cell phones, laptops, and other devices are actually recycled. The other 95 percent more than likely end up in landfills, leading to toxic escape of gases and leaching of cobalt, nickel, and other minerals. Perhaps this is why, as some lithium-ion car batteries are nearing the end of their useful lives, the European Union is taking steps to keep them out of landfills.

European carmakers are now responsible for collecting and recycling batteries, and a Belgian company, Umicore, is investing €25 million in an Antwerp pilot plant to recycle the batteries and to use a smelting process to recover cobalt and nickel ore for Tesla and Toyota. It's important to note, however, that lithium itself requires a much more expensive recovery process that may not be practical for at least a decade. What's more, the EPA cautions that nickel and cobalt (both also used in the production of lithium-ion batteries) represent significant additional environmental risk.

Researchers in China have developed a simpler, less energy-intense process to recycle and recover lithium cobalt oxide battery cathodes used in smartphones and laptops, along with the more complex lithium metal oxide materials used in EV batteries. The result is a marginally environmentally safer shortcut that creates new battery cells with the regenerated cathode material; these have the same energy storage capacity, charging time, and lifetime as the originals.[10]

Closer to home, companies in Bend, Oregon, and San Francisco, California, are also developing new processes to perform "direct cycling" of battery cathode materials. The jury is out as to whether these techniques will be adopted ubiquitously in the mining and resource recycling industries, but the logic is surely there, especially given that, by one account, the industry will have at least 18 million hybrids, plug-in hybrids, or pure electric vehicles using nickel–metal hydride or lithium-ion batteries by 2025.[11]

Myth: Lithium-ion batteries have no longevity.

REALITY

Tesla reports that its battery capacity remains at 95 percent at 50,000 miles, losing only another 5 percent in the next 150,000 miles. After 200,000 miles of driving, a Tesla battery can still have close to 90 percent of its capacity.[12] Even when batteries "expire," some companies are expecting to retrofit them for commercial or residential energy storage rather than recycling or deconstructing them for parts.

Myth: Lithium-ion batteries harm the environment.

REALITY

This one is complicated. The European Union reports that lithium-ion batteries, along with nickel–metal hydride batteries, consume a lot of energy, and 1.6 kilograms of oil is required per kilogram of battery produced. In addition, lithium-ion batteries ranked the worst in greenhouse gas emissions, with up to 12.5 kilograms of CO_2 equivalent emitted per kilogram of battery thanks to mining, processing, and the full manufacturing life cycle.[13]

What's more, lithium-ion mining can be environmentally unfriendly. Because lithium most commonly occurs in compounds such as lithium carbonate, chemical processing and large amounts of water are required to extract the pure lithium element. Toxic chemicals are used for leaching purposes.

Rare earth mining, which is more energy-intensive and generates more pollution than other types of mining, is also an issue: Up to 95 percent of the so-called rare earth elements (see page 4 in the introduction) are mined in China, which hasn't spent much time on pollution control measures. The rare earth extraction process requires the injection of ammonium sulfates and acid into the earth, leaving behind pools of toxic chemicals in water, including radioactive thorium (thorium, when ingested, produces leukemia and cancers of the pancreas and lungs), and poisons livestock, food, and, ultimately, farmers' ability to feed their families.[14]

Rare earths, or lanthanides, along with scandium and yttrium, are critical elements in the manufacturing of electric motors and magnetic components in EVs. A typical Toyota Prius, for example, uses 2.2 pounds of neodymium to manufacture each hybrid vehicle's electric motor. Almost all rare earth mining is done in China.

In Inner Mongolia, the town of Baotou, which produces two-thirds of the rare earth products used worldwide, has seen a huge lake and abundant fruit and vegetable farming area completely destroyed by toxic rare earth mining effluent.

In other parts of the world, however, companies are using cleaner techniques for extracting and processing rare earths. In Chile, a company called Biolantánidos is using biodegradable chemicals and planting trees once their extraction efforts are complete. The United States is exploring techniques for extracting rare earth mineral deposits from discarded coal using a milder ammonium sulfate chemical bath.

The Search for Perfect Chemistry

Where do we go from here? Companies and research institutions are working hard to find the ultimate solution to lithium-ion batteries. Most promising are metal air batteries (a metal at the anode; oxygen or other metals like nickel, cobalt, and manganese at the cathode), which show tremendous promise,[15] with varying energy densities, storage capacities, and speeds or recharge.

One of the front-runners is the energy-dense aluminum-air battery, which is used today principally for military applications. Aluminum is the most common recycled element on earth, so batteries of this genre could, with proper development, penetrate commercial markets abundantly and cheaply. In 2013, Israeli company Phinergy demonstrated a viable aluminum-air battery that can power an electric car more than 330 miles using a specialized cathode and potassium hydroxide. The battery anode, made of aluminum, required a replacement after 1,200 miles (2,000 kilometers).

Despite its potential, the aluminum-air battery comes with some significant drawbacks. For one, aluminum corrodes fairly rapidly during regular battery use. Plus, aluminum-air batteries aren't easily rechargeable.

In early 2015, Fuji Pigment Co. Ltd. announced that it has developed a new type of aluminum-air battery, using salt or normal water as an electrolyte and a modified structure to ensure a longer battery life. The battery's theoretical capacity was 40 times as energy-dense as lithium-ion, carrying a specific energy level of 8,100 watt-hour per kilogram compared with lithium-ion's specific energy levels within 100–200 watt-hour per kilogram.[16] Fuji Pigment is also experimenting with an ionic liquid electrolyte rather than saltwater. A TiO_2 (titanium dioxide) internal layer has been installed to separate the electrodes and accumulation of by-products. Ultimately, these or another aluminum-air upgrade could make batteries rechargeable.

The lithium-air or lithium-oxygen battery is another widely known experimental candidate. It pairs lithium and ambient oxygen in a non-aqueous (solid-state) battery. With a theoretical specific energy of 11,140 watt-hour per kilogram, this battery approaches the energy density of gasoline. In practice, the battery is five times more powerful than a commercial lithium-ion battery and sufficient to run an EV more than 310 miles on a single charge. But the battery requires additional chemical modifications and testing for commercial use. Other solid-state batteries are under development in Switzerland using sodium and magnesium crystalline structures without an aqueous electrolyte. Ions are displaced in a solid-state battery by developing crystalline structures that allow the ions literally to move through the electrolyte to the poles, releasing electrons and creating electricity.

No matter where you stand on the EV-versus-ICE debate, it's fascinating to see the wave of innovation and creativity that has accompanied the latest generation of EVs.

The U.S. Naval Research Laboratory, for one, teamed up with clean battery technology developer EnZinc to create an alternative to fire-prone lithium batteries that have been banned in some applications (including e-cigarettes) from U.S. Navy ships. This R&D partnership proved successful, resulting in a safe, high-performance nickel-zinc battery with a wide range of applications, from shipboard power for naval vessels to EVs and hybrids for the general public.[17]

EnZinc co-founder and company president Michael Burz, who worked on the design of the Tomahawk cruise missile and for Nissan, explained the battery's "spongy" design. "It looks like the sponge on your sink, but on a micro-scale—nanometers," Burz told *North Bay Business Journal*. "Zinc oxide forms on the outside skin of the sponge, but the inside walls of the sponge are clean. They carry current [in] a continuously wired structure. There's always a path for electricity to travel."[18]

Burz believes their battery will easily be able to compete with lead-acid batteries that have dominated the market: "We estimate that where a lead battery costs $130 per kilowatt hour, a zinc one will cost around $150—with twice to three times the energy and a much longer life and half the weight."[19]

What's more, EnZinc's product claims to be completely recyclable.

EnZinc is getting the battery production-ready for the market, according to its website, but (like most new breakthrough tech) we will likely be waiting a few more years before the production system is capable of producing them commercially.

In the meantime, other battery innovations are on the horizon. Stanford University professor Hongjie Dai and doctoral candidate Michael Angell invented a battery that uses urea—found in the urine of mammals and a common ingredient in fertilizers—as the electrolyte. "So essentially, what you have is a battery made with some of the cheapest and most abundant materials you can find on Earth," Dai said. "And it actually has good performance."[20]

Other researchers have announced batteries that use hydronium, glass, and even organic materials intended to function like an electric eel. It will be fascinating to see if they're successful.

In France, meanwhile, is EP Tender, one startup that is testing a range extender in the form of a small, rentable trailer that's about four feet long and weighs 400 pounds.[21] If it succeeds, the trailer should add about 500 kilometers (310 miles) to EVs' range, going a long way to ease the range anxiety felt by many long-haul EV drivers. EP Tender's goal is to make the trailers available throughout France via a network of 400 charging stations.

It's worth pointing out, though, that these trailers produce electricity with a combustion engine and alternator—and they need gasoline to function. Or, as Jasmina Schmidt wrote in a piece on EP Tender for *Reset*, "as soon as the trailer is added, the car becomes a kind of hybrid."[22]

But EP Tender's trailer is still a work in progress. The company says it's hoping to develop its technology further, so the trailer can be powered with fuel cells or a larger battery. "[They] argue that their range extender would only need to be used for a small number of trips each year, less than two percent of total car usage, so they claim. And if a solution like this is enough to get more electric cars on our roads, then that would already be a huge achievement," Schmidt said.

CHAPTER 13

ROADS:

HEY, ARE YOU GOING TO PAY FOR THAT?

Imagine, if you will, an election-style debate.

At one podium stands Ron Leys, a longtime participant in local government: "Because EV drivers in the United States don't buy gasoline, they aren't contributing anything to cover their use of public roads and are causing the decline of funds to pay for road upkeep."

At the other podium, urban farmer and community-builder Joshua Feyen disagrees. "It's not the EVs that are skewing the revenue for road maintenance," he says. "Drivers across the board—in EVs and ICEs alike—are buying less gas, as they're driving fewer miles in more efficient cars."

The debate continues, as each side presents the merits of its case. In the end, we're left with a pressing question: *Whose responsibility is it to take care of our public roadways?*

If you use public roads, shouldn't you help fund their upkeep? Whether you call it "your fair share" or a "use fee," a small amount from every driver is necessary to make sure that we can continue to use our public thoroughfares.

At least, our lawmakers think so. That's why we have a federal gasoline tax, as part of the price per gallon, that is funneled into an account to pay for road maintenance.

Gasoline Tax: Simple Brilliance or Outdated Scheme?

In 1956, the federal Highway Revenue Act introduced the Highway Trust Fund (HTF),[1] which mandated a tax of 3 cents per gallon to be used exclusively for highway construction and maintenance. This was slated to expire in 1972 but instead has been increased over the years. Just a few years after inception, the tax was increased to 4 cents. Then it jumped to 9 cents in 1983 (with 1 cent going into a new Mass Transit Account created by 1982 Surface Transportation Assistance Act). The Omnibus Budget Reconciliation Act of 1990 bumped it up to 14 cents, with 2.5 cents going toward deficit reduction. The Omnibus Budget Reconciliation Act of 1993 increased the gas tax to 18.4 cents, but all the increase went toward deficit reduction.

Here's the basic concept behind funding road building and maintenance via a gasoline tax: *The more you drive, the more wear and tear your vehicle does to roads and bridges—and the more you pay at the pump, as you use more fuel, the more you contribute to the maintenance fund. If you drive a bigger and heavier vehicle that inherently does more damage to roads, you burn more gas and pay a larger amount into the fund by buying more fuel.*

The gasoline tax created a "use fee" that was easy to collect. This simple-yet-effective system of taxing drivers at the pump effectively charged all road users their "fair share" and funded almost all the needed road maintenance, repair, and building costs around the nation via the federal Highway Trust Fund.

And at pennies on the dollar, most drivers didn't mind (or didn't even notice) that they were paying into the system.

The average driver doesn't really think about this gasoline tax, said

Leys. "When people put gas in the tank, they're not aware that they're also paying a federal tax. It's not like sales tax. When you buy a TV, the clerk says, 'That's $500 plus $25 in sales tax.' You're very aware. That connection isn't made for fuel tax. It's a hidden tax."

Unfortunately, this system has proven a bit inadequate of late.

Road maintenance is an expensive and ongoing endeavor, and the government has not raised the gasoline tax rate in 25 years. Since then, the percentages of gasoline revenue have been reallocated many times— the 1997 Taxpayer Relief redirected the 1993 increase back into the Highway Trust Fund, and the American Jobs Creation Act of 2004 eliminated a partial exemption on gasohol that was part of a 1978 alternative fuel incentive—but the tax rate itself didn't actually increase.

According to the U.S. Department of Transportation Federal Highway Administration (FHWA), "At the close of FY 2015, the Highway Account held a cash balance of $9 billion. At the same time, though, there were $64 billion in unpaid commitments against the HTF: authorizations that FHWA had previously apportioned or allocated to States (and other eligible recipients), but for which the Treasury Department had yet to outlay cash."[2]

Changing Times Require Changing Rules

This gas tax scheme seemed like a genius solution for the time. Everyone was contributing, and their contributions were directly related to how much they were driving. But it was established back in a different era, when (a) many drivers were logging lots of miles, and (b) a good portion of the cars on the road were gas-guzzlers. Here is where we are now running into problems: Both trends are changing—dramatically.

First, let's consider driving habits.

In general, Americans are driving less. In recent years, the number of miles driven in the United States was stagnant or generally declined. Baby boomers—until recently our largest living generation—began aging out of their driver's licenses, while millennials—whose ranks among our population now exceed the boomers—began foregoing learning to drive entirely. At roughly 75 million and 83 million people, respectively,[3] that's a significant chunk of the eligible population who weren't driving at all.

A 2016 study from the University of Michigan Transportation Research Institute shows that the past several decades have seen a sharp decline in the percentage of young people getting their driver's licenses (see figure 13.1). In 1983, almost half of 16-year-olds were licensed drivers; three decades later, that number was less than a quarter. Even older drivers, who had a sharp increase in the 1980s, started a similar decline.[4] Although it appears that driver numbers might have hit rock bottom in 2014, the rebound that the U.S. Department of Transportation reported in January 2019 is anything but dramatic.[5]

Age	1983	2008	2011	2014	2017
16	46.2	31.1	27.5	24.5	26.3
17	68.9	50.0	45.0	44.9	46.9
18	80.4	65.4	60.3	60.1	62.1
19	87.3	75.5	69.3	69.0	71.6
20–24	91.8	82.0	79.7	76.7	79.6
25–29	95.6	86.3	87.5	85.1	85.9
30–34	96.5	90.6	89.1	86.6	89.9
35–39	94.9	91.7	90.2	87.9	90.4
40–44	92.2	91.9	91.6	89.1	91.1
45–49	92.5	93.0	91.9	90.5	92.2
50–54	91.4	94.2	92.2	91.2	92.7
55–59	88.2	94.9	93.2	91.8	93.2
60–64	83.8	95.9	92.7	92.1	93.4
65–69	79.2	94.0	93.0	91.4	92.8

Figure 13.1. Percentage of people getting driver's licenses by age.

Younger people, in particular, were opting to live in neighborhoods where they could walk or bike or take mass transit to work. As Feyen said, "Millennials eschew the traditional one-car-per-person household, opting to share cars, carpool, rent, or use community cars as needed."

Not only were there fewer drivers in these categories, but the trend among those who are still driving was to drive less as well. Across the country, eligible drivers were becoming less likely to even *own* a car. The University of Michigan research found that 8.7 percent of U.S. households owned no vehicle in 2007; in 2012, that statistic was 9.2 percent. And the swing away from car ownership was, somewhat expectedly, more dramatic in larger U.S. cities with good public transportation systems (see figure 13.2).

City	% Households
New York City	56%
Washington, DC	38%
Boston	37%
Philadelphia	33%
San Francisco	31%
Baltimore	31%
Chicago	28%
Detroit	26%

Figure 13.2. Percentage of car-free households in 2012.

Now that we're seeing growth in our economy once again, the good news is that those numbers are finally rising: The Federal Highway Administration estimated that 2016 was the fifth straight year of increased mileage on our public roadways. But the bad news is that this same increase "underscores the demands facing America's roads and bridges and reaffirms calls for greater investment in surface transportation infrastructure."[6]

And Leys knows firsthand just how detrimental these demands are. He spent many years as an alderman for the city of Prairie du Chien, Wisconsin. In that capacity, he helped the city decide which roads to work on and which to avoid.

"The roads in Wisconsin were once in good condition. It's important for our tourism. But that began to erode in recent years, especially ten years ago when we stopped indexing fuel tax to inflation," he said. "That put a lid on the fuel taxes. Since then, the real dollars available for road construction and repair have been going down."

Every year, Leys said, the Town Board would decide which roads were in the worst shape and then choose just a few to sealcoat—which is simply a layer of thick gooey oil sprinkled with stones and steamrolled together. "This is a very cheap form of pavement, but it was all we could afford. The state regards this as temporary: It might last five years, then you just get more potholes. And now we have some of the worst roads in the state," he said.

But it's not just the amount we're driving (or *not* driving) on these poor roads. It's also the cars we are choosing (and *not* choosing) to drive.

In general, today's cars are much more fuel efficient. That's not to say that big pickups and SUVs aren't prevalent or that every driver gets at least 50 miles per gallon, but automotive engineering is making huge strides. Back when the tax scheme was designed, there was no way for anyone—auto manufacturers, drivers, or legislators—to conceive of today's mileage abilities.

Over the years, a stream of initiatives has encouraged this technological growth targeted at automotive fuel economy. Corporate Average Fuel Economy (CAFE) standards have been around since 1975, way back when the Highway Trust Fund gasoline tax was just a few cents per gallon.

"The purpose of CAFE is to reduce energy consumption by increasing the fuel economy of cars and light trucks," the Department of Transportation explains. "When these standards are raised, automakers respond by creating a more fuel-efficient fleet, which improves our nation's energy security and saves consumers money at the pump."[7]

The Energy Tax Act of 1978 subsequently imposed the aptly named "Gas Guzzler Tax" to openly discourage the manufacture and purchase of fuel-inefficient vehicles. With certain exemptions, carmakers had to produce cars that met a minimum fuel economy level of 22.5 mpg or face the tax penalty based on actual mileage—and these ranged from $1,000 to $7,700 for *each* vehicle.[8]

More recently, the Department of Transportation and the Environmental Protection Agency established federal rules in 2010 to increase the fuel economy of all new passenger cars and light trucks sold in the United States. These increasingly stringent standards, applicable to all new 2012 through 2016 model-year vehicles, required car manufacturers to improve fuel economy by 5 percent every year to the ultimate goal of 34.1 mpg for 2016s.[9] Two years later, the standards extended to model years 2017 through 2021, requiring 2021s to average 40.3–41.0 mpg.[10]

And carmakers have heeded the call for these types of fuel economy mandates. They introduced smaller and lighter cars made of aluminum or composite, with more aerodynamics and less drag coefficient. They also said farewell to the carburetor, the clever mechanical device (that's been around pretty much since the dawn of the automobile) that mixes the correct proportions of fuel and air to power the car. Apart from motorcycles and race cars, production vehicles now use fuel injection to feed fuel and air into the engine's combustion chamber. This system of direct fuel injection improved engine efficiency—and, in turn, increased cars' fuel economy—far beyond the capabilities of the carburetor.

Car manufacturers have also integrated technology for even greater engine efficiency. Computers help mix the air and gas in a more reliable way, dictate and monitor the fuel injection, and even monitor the air in the tires to prevent incorrect air pressure that can drastically alter the miles per gallon achieved.

Our collective need for less gasoline also led to a drop in the number of gas stations around the country. In fact, U.S. Census Bureau statistics note that, between 1997 and 2012, almost 10 percent of the nation's gas stations closed.[11] This also means that there are fewer places for drivers to contribute via gasoline tax.

The end result? Less fuel purchased means less gas tax collected, leaving the coffers of road maintenance depleted. Which is to say, the current gas tax scheme has gone awry.

A Free Ride? Or a Fair Ride?

Meanwhile, back at Debate Central, the parley might continue something like this:

Leys argues, "EV drivers need to start paying their way—somehow. I've heard complaints already: 'I have an EV, I'm contributing much less to air pollution, I shouldn't be penalized.' But they are freeloaders. Since they're not paying through gas taxes, they need to make it up somewhere."

Feyen counters, "Even though EVs do not burn gasoline and so do not contribute to the gas tax, they might actually contribute *more* gross revenue and *more* revenue per mile driven to the Highway Trust Fund through higher registration fees and sales taxes because of their higher sales prices."

Leys says, "With more and more electric vehicles on the road every year, the deficiencies of the Fund are growing. The problem is simply compounding itself."

Feyen responds, "You can't blame EVs for the significant loss in gas tax revenue—they still account for less than 1 percent of the nation's fleet. At the end of 2015, they comprised a meager 0.6 percent of all U.S. light-duty vehicle sales.[12] There are simply not enough EVs driving around today to make a difference to the Fund's revenues."

Despite their opposing viewpoints, both sides agree that we desperately need to introduce new measures to ensure that our public roadways can be maintained. But the experience on the Highway Trust Fund proves that it's going to take more than just mandates to bring back a functioning system.

"We're still doing things like we did in the 1920s, as though nothing has changed in the world. It's time for some new thinking," said Leys.

So what would the system of "fair" taxation look like?

The Need for New Legislation

For the past 20 years, politicians on both sides of the aisle have wrangled with this question. Individual members of Congress have proposed suspending the federal gas tax or repealing it altogether, each without success. The issue took center stage during the 2008 presidential campaign, with Senator John McCain (R-AZ) calling for a suspension during the peak summer driving season. Democratic rivals Senator Hillary Clinton and Senator Barack Obama clashed, as Clinton agreed with McCain and Obama opposed this suspension.

In June 2015, Joseph Kile, assistant director for microeconomic studies at the Congressional Budget Office, testified before the Senate

Committee on Finance, stating, "The Highway Trust Fund cannot support spending at the current rate . . . if nothing changes, the trust fund's balance will be insufficient to meet all of its obligations in fiscal year 2016, and the trust fund will incur steadily accumulating shortfalls in subsequent years."

He urged Congress to take definitive corrective action instead of relying on handouts from other areas in the government coffers, warning that "borrowing is only a mechanism for making future tax revenues or user fee revenues available to pay for projects sooner; it is not a new source of revenues. Borrowing can augment the funds available for highway projects, but revenues that are committed for repaying borrowed funds will be unavailable to pay for new transportation projects or other government spending in the future."[13]

Leys agrees that borrowing is a poor alternative. "We don't have enough money to pay for our roads, so they keep talking about borrowing money to cover those costs," he says. "I don't mind borrowing money for a one-time expense to be enjoyed for many generations, but the government should never borrow for a current expense. You should just tax for current expense. It's just taking from the next generation."

However, in the intervening years, legislators have come to no conclusions about how to breathe life back into our failing Highway Trust Fund.

And since fiscal 2008—at the same time McCain, Clinton, and Obama were arguing about it—the fund's expenditures have exceeded its revenues in each subsequent fiscal year to date. Because they could not simply update the system already in place to call on each driver to chip in just a little bit more for each gallon, lawmakers were forced to seek alternative methods for funding the trust. General revenue funds have been poured in to cover the shortage to the tune of $143 billion since 2008.[14]

And since then a variety of solutions have been proposed to improve our antiquated gasoline tax scheme. Let's look at a few of them in more detail.

1. WE COULD CONTINUE TO EMPLOY THE CURRENT GASOLINE TAX.

This system is already in place, and it's a small enough expense that drivers hardly notice. Even a jump in the percentage would likely get little more than a passing consideration. Leys points out that fuel taxes in Europe are very high, but those democracies don't vote out the politicians over those rates. Because wear and tear is directly connected to the speed of the auto, Germany imposes a special tax on expensive high-speed autos, and people there just seem to accept it.

In fact, Leys said that this period of extreme gas price volatility is probably the best time to raise the fuel tax. People wouldn't notice a nickel or dime. They're acutely aware of the relationship between taxes and schools because they pay property taxes all at once, and that's a big check to write. But the relationship between taxes and roads is much more elusive.

"Drivers grumble about the roads but don't make the connection," Leys says. "I don't hear anyone saying, 'Please raise my taxes so I can have better roads.' And the legislators don't want to have to admit, 'I raised your taxes.' They wanted to say that taxes didn't go up on their watch."

And because the rate hasn't been increased in over 25 years, the jump would need to be pretty severe to make a dent in the Highway Trust Fund's shortfalls. If tax rates had been indexed for inflation since 1993, the current tax on gasoline would be up to about 31 cents per gallon.[15]

2. WE COULD USE TOLL ROADS TO RAISE FUNDS.

Tolls help cover road funding, and they make it easy to charge drivers based on how much they're actually driving. But it may be hard to institute tolls on existing roads.

What's more, Leys added, if it's about fairness, toll roads don't really fit the bill.

"I'm opposed to toll roads—they are just a scheme to tax someone else to pay for your roads," he said. "Toll roads are always set up so that they pay for more than the use of the road. That money goes for road repair elsewhere in the state. People who drive long distances end up paying to maintain other roads they're not even using."

He compared toll roads to the Panama Canal: Passage through it costs up to $1 million, and the country makes a huge profit, which it uses to pay for everything else in the country.

3. WE COULD ELIMINATE THE GASOLINE TAX AT THE FEDERAL LEVEL.

Many decry the waste of high government spending that comes from federal involvement, maintaining that reducing federal involvement outside of the interstate highway would hand road funding back to the states. This has led to measures such as the Transportation Empowerment Act, a bicameral bill that has been introduced several times in recent years to gradually reduce the federal gas tax and give individual states the control of transportation decisions inside their state lines, from what to build and how to build it to how much tax to charge for gasoline.

The thinking behind this is, in addition to eliminating a layer of federal bureaucracy that could save time and money, states know better than Washington bureaucrats which road projects are the most urgent concerns.

"Our streets are in horrible shape, but we don't have any money to deal with it," said Leys. "The legislature has taken away the power of the local community. They don't trust the people in the local communities, and I've seen that's true under both Republican and Democratic leadership."

"Give us people in local government the power to decide," Leys said. "The voters know where we live."

Several states have figured out how to do just that, while others are considering new ways. Each individual state already determines its own annual vehicle registration rates and fee structure:

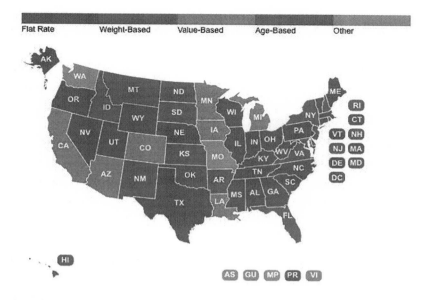

Figure 13.3. Registration and title fees by state.
Source: National Council of State Legislatures, http://www.ncsl.org/research
/transportation/registration-and-title-fees-by-state.aspx.

Oregon started us off with a voluntary "usage tax" program in 2013. Using either a GPS device or odometer readings, owners of cars that get more than 55 miles per gallon are charged a flat annual fee of $542.50, or a usage fee of 1.55 cents per mile. Gas tax paid at the pump is refunded by the state to avoid double taxation.

Shortly after the Leys interview, the Wisconsin legislature added an "EV Surcharge" to annual vehicle registration. A yearly license plate sticker will now cost $175 for electric cars (defined as those propelled solely by electrical energy and not capable of using gasoline, diesel fuel, or alternative fuel), while drivers of ICE cars will continue to pay the standard $75. As Leys put it, "EV drivers are now paying $100 per year for the privilege of going electric."

Several other states have also introduced measures to boost revenue from drivers, from adding an "electric vehicle fee" to implementing weight-based tariffs. Alabama, for example, charges an additional $50 fee for passenger vehicles over 8,000 pounds. California adds a "Transportation Improvement Fee" based on vehicle value. Colorado's license fees are based on the age and weight of the vehicle, with an additional $50 assessed for EVs.

But it seems that, at present, these state programs need to work hand in hand with a federal system to bring our road funds back up to required levels.

4. WE COULD REPLACE THE GAS TAX WITH A "USE TAX."

This would be based on the actual number of miles each vehicle logs—which seems like the most sensible alternative for many reasons. This per-mile tax would be on top of the annual registration fee that each state already has.

"It could be an annual fee, but many people would balk at a huge bill once a year," said Leys. "Monthly might be easier to stomach."

Taxing people per mile driven would not be hard to do, Leys said. Opponents might be concerned that tracking drivers' miles constitutes an invasion of privacy, but the implementation would be relatively easy and function just like the electric company.

"Electricity customers are taxed per kilowatt hour for the electricity they use. It used to be that someone had to go read all the meters each month; now it's done automatically through a signal that broadcasts from each home to a meter reader," Leys said. "We have the technology to do that for cars, too. The car could automatically record miles through the odometer, and an instrument could automatically broadcast that number to a central office. Just think: It's not too different from the system of city toll roads, like the iPass system—we don't even have to stop to pay tolls anymore, or even slow down!"

And here's where our two debaters actually converge.

Leys likes the "use tax" option because it allows for accountability and might open our eyes to an important relationship: "People do understand that we have to pay for our roads, but in a vague sort of way. If drivers got a regular bill, they would connect that to the roads in the same way people connect property tax to schools."

Feyen likes the fact that it seems fair to everyone: "I like this because it's a program that doesn't discriminate between electric vehicles and efficient conventional ones, and comes closer to charging for actual use of roads and bridges."

What's more, this type of system might have the added benefit of encouraging more people to take public transit.

"If drivers actually noticed the tax, they might decide to hop on the bus or even bike to work more often to reduce their taxes," said Leys.

But even if it doesn't change behaviors, billing drivers for each mile driven is fair. It doesn't "penalize" anyone for the type of car they drive, and it funnels much-needed money into the maintenance funds to fix the roads we're all using.

The Issue of Fairness

"I'm a car guy. I'm eighty-one years old, and I've owned many cars over the years. I bought my Chevy Volt in 2012. This is the best car I've ever had by far—I would not hesitate to buy another one," said Leys, who added that his car "runs on sunshine."

"I have solar panels on my house," he said. "The guy who installed my 'solar system' looked at our electrical bills for a year and installed enough to provide that much electricity. We buy electricity in the winter, and in the summer, we power our whole house and have enough to sell back. I am running this car on sunshine, paying no fuel tax."

Leys said that while he enjoys the fact that he's not burning gasoline, he also feels a bit guilty: His experience with the city council has made him acutely aware of the problem of building and maintaining the roads.

Something, he said, needs to be done.

"I've been lobbying our legislature, through my one-man campaign, since 2012," he said. "I've been telling them, 'You've gotta raise my taxes!' I guess they finally listened when they set up the new system for 2018. I think it's fair—I figure that's about what I would have paid in gasoline anyway."

Leys believes that a per-mile tax could be sold to the public now, if lawmakers took things slowly and started with plenty of public discussion, hearings, and education. People need time to think about it and get used to the idea of doing things a little differently.

He is waiting with bated breath to see how the EV drivers of Wisconsin will react when they receive their annual vehicle registration renewal. In his worst-case scenario, people will be less inclined to buy electric cars. Ultimately, he says, it will all come down to whether the public believes that the fee is fair.

"The idea of 'fairness' is a very strong one in people's brains," he said. "If people decide that something's not fair, you'll have a hard time convincing them otherwise."

EVS INTRODUCE NEW ETIQUETTE ON SHARING

As the number of EVs on the road continues to grow, their drivers are going to need to remember some advice from their kindergarten teachers: *You need to share.*

"I want to pay my fair share," Feyen said, noting that he likes to leave a couple of bucks for the homeowner whenever he charges at someone's house. "I don't expect anyone to pay for my drive."

But Feyen charges his EV for free at work—not uncommon nowadays for many businesses. And that leaves him a little uneasy, for a couple of reasons.

"My employer shouldn't fund my commute—they don't fund other people's commutes. But, for now, they're trying to encourage people to buy EVs. It's a 'corporate responsibility' thing. I'm not complaining—it's nice. But eventually more EVs are going to start arriving in the parking lot, too many for the company to keep subsidizing, for lack of a better word. And there needs to be a system in place before that happens."

In addition to the concept of paying his "fair share," Feyen sees another, perhaps more important, reason for the fees at public charging stations: encouraging a new kind of driver etiquette. In short, EV drivers need to remember to share.

continued

"Imposing a time-based fee gets people out of those spots, so other drivers can charge, too. Otherwise you could have fully charged cars just taking up those spaces and blocking others from charging."

Sure, with today's relatively low numbers of EVs on the road, it's easy to forget fellow chargers. But it's a bad habit to assume no one else might come along. A fee-based charging system helps remind them to be conscientious of their fellow chargers . . . so if the day ever comes where EVs are lined up outside the mall waiting to charge, those using the stations will remember to pause their shopping to move their cars.

Feyen explained that a "charging etiquette" at his company developed organically.

"We are currently 15 or 16 EVs out there now. We coordinate with each other through a messaging system—when we're leaving, if we need a charge right away, that kind of thing. In the early days, we just all adopted a sort of universal 'courtesy system' that if you pull in at 8 a.m., you pull out around noon."

And now the company has two spots per charging station, as a direct result of the number of EVs in the lot.

"The EV drivers have definitely influenced the company, not the other way around. We got them to designate a second slot next to each charge station, so we can pull in to wait for the one who's charging and not have to swap spots. This is infrastructure change that WE caused."

Many EV models have an alarm feature if they get unplugged from the charger. Feyen said that all the drivers at his office have turned off their alarms to make it possible for one driver to unplug someone else.

Because many public charging stations—including those at Feyen's office—put the chargers in premium parking spots, there's a perception there that EVs are getting preferential treatment. And perception is reality.

"I want it to be utilitarian: I'm just here to charge, not to get the premium parking spot. This could be a PR function, something to foster goodwill and positive community relations. But I say, especially at work, put them in the back!"

CHAPTER 14

PLASTICS AND FRIENDS:

THE MARRIAGE OF EVS AND O&G

Of all the great modern partnerships, few have endured as long as the one between the automotive and oil and natural gas industries. The relationship, now well into its second century, is as symbiotic as oxpeckers landing on zebras to pick off parasites—the bird enjoys a free lunch while the animal gets a little no-cost pest control—or bees pollinating flowers. The benefits are mutual.

It's been that way for more than a century, ever since the first cars rolled off Henry Ford's assembly line and oil companies were there, gasoline cans in hand.

But gasoline wasn't on the production schedule when the first oil refineries set up shop at the end of the nineteenth century.

The fact is, those early oil refineries hadn't set out to make gasoline at all. Buoyed by a spate of oil strikes at the end of the nineteenth century, they had hoped instead to cash in on the home lighting business, making kerosene to replace whale oil in lamps. There were three, equally convincing arguments in favor of kerosene: It was every bit as bright as whale oil, cheaper, and less likely to explode. (That last point was enough to make most people converts.)

As for gasoline, it was considered nothing more than a vile, highly flammable by-product of the kerosene distillation process with no apparent commercial use. In an ecologically disastrous attempt to dispose of the stuff, barrels of gasoline were regularly dumped in waterways near refineries. Speculation was that the Cuyahoga River, for one, was so polluted with gasoline that a coal dropped from a passing steamboat would set the water ablaze. (This prophecy came to pass much later, in 1969, when oil-soaked debris on the Cuyahoga was sparked by a passing train.[1] The event is widely considered a primary catalyst for the Clean Water Act of 1972.[2])

That all changed with the waste-averse tycoon-to-be John D. Rockefeller, the Standard Oil founder who'd mastered penny-pinching during his underprivileged upbringing. Never one to discard something that could eventually be put to good use—and always interested in new ways to grow what would become his empire—Rockefeller had a brighter idea: Rather than tossing out gasoline like trash, he would harness the would-be fire hazard to use as fuel for the refining process. Once gasoline was proven effective in refinery equipment, attention was turned to finding other applications. Soon, Standard Oil—and several other companies who'd piggy-backed on Rockefeller's idea—began promoting gasoline to the auto industry, which was looking for a reliable way to quite literally move cars. If the fates hadn't collided exactly as they did—if the mass production of automobiles powered by internal combustion engines hadn't occurred just as Rockefeller was looking to, in today's terms, monetize gasoline—all these years we might have been tooling around in cars powered by kerosene, steam, or primitive electric motors. Instead, all over the world, we rely on vehicles that burn petroleum-based fuels. That fact is no more pronounced than in the United States, home to less than 5 percent of the world's population but more than 20 percent of the world's autos.[3]

For a perspective on the importance of gasoline to American life, consider this: It's the most-consumed petroleum product in this country.

The Energy Information Administration estimated that, in 2017, we used an average of 392 million gallons per day—or about 47 percent of total U.S. petroleum consumption.[4]

Clearly cars and petroleum have a thing going on. Can anyone really tear it asunder?

Without Oil and Gas, You're Going Nowhere

Nicolas Hulot is a green activist and France's ecology minister. In June 2017, he made a bold declaration bidding adieu to gas and diesel cars in his country—a statement made even more stunning given that French auto production is as much a point of pride as an economic engine. By 2040, he said, the sale of cars running on petrol or diesel would be outlawed.[5]

His remarks were followed a month later by a similar, but more tempered, pledge out of Britain. Jesse Norman, the undersecretary at the Department for Transport, said, "The Government has a manifesto commitment for almost all cars and vans on our roads to be zero emission by 2050. We believe this would necessitate all new cars and vans being zero emission vehicles by 2040."[6]

Norway, China, and India crow that they, too, are waving goodbye to cars with internal combustion engines. Norway is on the leading edge of EV adoption; China is pushing plans for an annual EV production quota; and New Delhi has launched what it calls an aspirational plan that, by 2030, every vehicle sold in the country will be powered by electricity. The nation is hedging its bets by suggesting it will allow for "the logic of markets" to prevail,[7] which seems wise given the fact that large swaths of India aren't even on the grid.

With those bans in place and projections for EVs soaring, has the gas-automobile relationship started to turn parasitic?

Hold on a minute! Even if we lived in the kind of all-electric Utopia that occupies Elon Musk's dreams and gasoline were obsolete, the automotive industry would still rely on petroleum-based products, and plenty of them.

The New Car Diet: Heavy on Lighter-Weight Plastics

There's a reason we "put the pedal to the metal." Since the first car was introduced in 1918, steel[8]—an alloy made from iron metal—has been the dominant material in automobile manufacturing.[9] It's also the heaviest, making up 60 percent of the average car's weight. Historically, steel has been the preferred option for structural components requiring high strength.[10] But with the shift to lower- and zero-emission cars and the demand for increased fuel (or range) efficiency, manufacturers have been scouring the design labs for ways to produce lighter cars without sacrificing safety. One avenue is to incorporate more lightweight materials, including aluminum and plastic. There's even a movement afoot to make those bulky EV batteries smaller and cheaper—while maintaining range—and it depends on lighter weight plastic.[11]

Currently, there are about 260 pounds of plastic in the average car.[12] As more CO_2 reduction mandates come on board, that figure will only continue to rise, says Aafko Schanssema of PlasticsEurope, an association of plastics manufacturers based in Belgium.

The Canadian Plastics Industry Association agrees. According to a media release, plastics make up 50 percent of the volume of new cars but account for only 10–12 percent of the weight, helping make cars lighter and more fuel efficient. For every 10 percent reduction in weight, they add, fuel economy improves by 5–7 percent, reducing the cost of fill-ups and resulting in fewer CO_2 emissions.

Modern plastics are so versatile you can spot them almost anywhere, from bumpers and body panels to headlamps, windshields, and upholstery. Plastic fuel lines and tanks are less prone to corrosion than their steel counterparts while plastic housings protect the power trains, batteries, and electronics from corrosive elements, heat, and environmental damage.[13]

To borrow from the 1967 film *The Graduate*, "There's a great future for plastics." And that's especially true for the emerging EV market: The compounded annual growth forecast for global plastic in EVs is 37.3 percent, reaching $1.49 billion by 2021.[14]

PLASTICS: STILL A GREAT FUTURE

With the demand for high-quality plastics growing all over the map—more products and new markets—and some forecasters predicting shrinking gasoline consumption beginning in 2026, some pretty big names in the oil and gas industry are revving up petrochemical production to keep their balance sheets in shape.[15]

You may already understand the link between petroleum, petrochemicals, and plastic, but for the uninitiated, a key ingredient in plastic is polyethylene. Follow the etymological breadcrumbs, and it becomes apparent that polyethylene is made from ethylene—which just happens to be the most important chemical manufactured in the world, at least when measured by tonnage. And where does ethylene come from? From crude oil and natural gas, through a process called "cracking" that uses heat and pressure to break larger molecules into smaller ones.[16]

It takes roughly 0.4 gallons of crude oil to make one pound of plastic. Globally, around 8 percent of the oil that comes out of the ground goes into plastics manufacturing.[17] Of that, 23,000 tons were expected to be used in 2017 in cars and trucks.[18]

But that just scratches the surface when you're performing the calculus of opportunity for the oil and gas industry. That's because crude oil is just one of the feedstocks for plastics—and not even the most important one, at that. According to the U.S. Energy Information Administration, more plastics are made from natural gas and more feedstocks are derived from natural gas processing than from oil.[19]

For proof of how attuned oil and gas companies are to plastic's growing potential, consider that Chevron Phillips Chemical Company is investing $6 billion in building a large-scale ethane cracker and two polyethylene plastics units to expand production capacity.[20] Although the Houston-area project was delayed slightly when Hurricane Harvey inundated the region, first phase commissioning took place in September 2017. The project came online in March 2018; the three plants together will be capable of producing a million tons of polyethylene per year.

Nearby, ExxonMobil is also expanding its polyethylene capacity from 1.2 million to 2.5 million tons per year as part of its $20 billion Growing the Gulf initiative.[21] Production from the first of two new 650,000 tons-per-year high-performance polyethylene lines kicked off in October 2017. In a statement, ExxonMobil said, "These performance polyethylene products will deliver significant sustainability benefits enabling lighter weight higher performance packaging, lower energy consumption and reduced emissions."[22]

And isn't that last point what EVs are all about?

How We Roll: Tires Are Made of Petroleum, Too

If you've ever kicked a tire, you probably realize you've scuffed your toe against natural rubber, which comes, unsurprisingly, from the sap of the rubber tree. But natural rubber is just one of the raw materials used in tire manufacturing. Most tires also consist of synthetically formed rubbers with the kind of complicated names apropos of an AP chemistry test: *styrene-butadiene rubber* . . . *polybutadiene rubber* . . . *butyl rubber*. Besides being tricky to spell, difficult to pronounce, and possible fodder for autocorrect hilarity, what these three synthetic rubbers have in common is that they are made using polymers, the chemicals found in crude oil.

According to the Rubber Manufacturer's Association, it takes about seven gallons of oil to create a standard car tire. Five of those are used as feedstock, while the other two are consumed as part of the manufacturing process itself.[23]

But synthetic rubber isn't the only place where petroleum is part of the tire-making process. It's also the key—or, more precisely, the only—ingredient in carbon black, the soot-like reinforcing agent responsible for a tire's strength and durability.

In a 2010 article, *Scientific American* magazine discussed the complexity of tire chemistry. In an interview, James Rancourt, a consulting polymer scientist who heads Polymer Solutions in Blacksburg, Virginia, told the publication that the tread compounds of a conventional tire contain equal parts natural rubber, synthetic rubber, and carbon black filler, which is created when crude oil or natural gas undergoes a controlled burn with a limited amount of oxygen.[24] It's interesting to note that until

1912, when B.F. Goodrich introduced carbon black into the vulcanization process—the method by which natural rubber is beefed up by the addition of sulfur or other additives or curatives—car tires were white.[25] As the automotive industry grew, so did the use of carbon black.

CARBON CRAYONS

Equally fascinating is the fact that we have the makers of Crayola Crayons to thank for carbon black in the first place. It rose from the search by partners Edwin Binney and C. Harold Smith for a pigment that could be used in permanent markers for crates and barrels. The pair, who had already invented dustless chalk, pioneered carbon black using petroleum from the booming Pennsylvania oil fields as feedstock. Mixed with paraffin, another oil field by-product, and wrapped in wax paper, carbon black became the first crayon. It was a commercial success, although it wasn't safe for use by children. A tinkering of the recipe allowed Binney and Smith to create the sixteen-pack of colors everyone loved as kids.[26]

From Under the Hood to Out on the Road

Of course, this chapter has only scratched the surface of the deep, complicated marriage of oil and automobiles. Look under the hood of any car—regardless of make, model, or fuel source—and you'll find that they all need lubricants to operate at peak form. The slippery stuff delivers a multitude of motor maintenance benefits: It reduces friction, heat, and wear; protects metal parts from corrosion; neutralizes acids that could build up in the engine; traps particulates that could damage parts; and prevents the accumulation of sludge.

Conventional cars with internal combustion engines use oil, as do hybrids (HEVs) and plug-in hybrid electric vehicles (PHEVs). In fact, compared to conventional ICE vehicles, HEVs/PHEVs require additional, higher performance grade lubricants. Pure battery-electric vehicles don't use oil, but they still rely on a small amount of greases and related secondary products to keep them performing efficiently.

Typically, lubricants contain 90 percent base oil (most often petroleum fractions, called mineral oils) and less than 10 percent additives. But even synthetic oils can trace their heritage back to the oil field—they're made from hydrocarbon-based polyglycols or ester oils.[27]

Batteries—yes, even those that power hybrids and EVs—also have some petrochemical-based components in them. The new generation lithium-polymer batteries that are smaller, lighter, and cooler? Substitute "plastic" for "polymer"—after all, the terms are basically interchangeable—and you'll see that, for all vehicles, petroleum is, quite literally, still in the running.

And don't forget about roads. There are more than 2 million miles of them in the United States and upward of 40,000 miles are added each year.[28] Nearly all of them—94 percent, to be precise[29]—are made of asphalt, a petroleum-based product that is strong, versatile, and weather- and chemical-resistant. Even though asphalt comes from the bottom of the fractional distillation process,[30] it is unrivaled at creating tough, durable surfaces—without it, our roads would still be a primitive, kicked-up mess of crushed stone and gravel.

Like gasoline before Rockefeller, asphalt has humble beginnings: It was a basic residue from the crude oil refining process. But it has been used to pave roads since at least 1870, when American city planners sought an economic alternative to imported bitumen. By 1902, companies in the oil hot spots of Texas and Pennsylvania were churning out asphalt; the following year, Congress established laboratories to test road materials and safeguard drivers. According to the American Oil & Gas Historical Society (AOGHS), within a decade, petroleum asphalt dominated the marketplace—and its prominence never wavered.

As the AOGHS notes, "The abundance of asphalt makes it seem unremarkable, yet without this basic residue from the petroleum refining process, bad roads may well have delayed much of the nation's economic progress of the twentieth century."[31]

'Til Death Do Us Part

The rise of the EV clearly doesn't equal the demise of the oil and gas industry. As Andrew Ward, energy editor for the *Financial Times*, wrote, there are plenty of opportunities "for oil companies to chase growth in other parts of the energy sector. Most large oil groups are investing heavily in natural gas in the belief that its relatively 'clean' characteristics, compared with coal and oil, will keep it growing longer."[32]

Others are participating in renewables: Statoil has interests in several offshore wind farms and is testing a floating wind turbine. BP is expanding its wind-energy portfolio. And some are cementing their relationship with the automotive industry by supplying two things the EV market can't grow without: batteries and charging stations. Total, for example, has purchased EV battery maker Saft and is considering installing EV charging stations at fuel stations in France.[33] In the United Kingdom, Shell is opening EV charging points at ten gas stations in London, Surrey, and Derby, where drivers will be able to recharge 80 percent of their battery in half an hour. This comes after Shell purchased the Dutch company NewMotion, which has 30,000 private charging points at homes and offices in Europe.[34] Shell is also committed to adding charging stations at locations in the Netherlands.[35]

When internal combustion engines came onto the scene, they created the market for gasoline and spurred increased oil exploration. And even today, as EVs make their way into the mainstream, the need for oil and natural gas hasn't gone away—and it's unlikely that the industry will go the way of whale oil in the future. This is especially true when you consider that ICE technology will more than likely become more advanced and produce less emissions—perhaps putting them on par, environmentally speaking, with EVs. Perhaps one day they'll even surpass EV technology. As more efficient gas-powered vehicles come online, that increased quantity will lead to greater demand for oil and gas.

Between their expansion into EV supplies and the conventional uses for petroleum products in vehicles and roads, the long-lasting, symbiotic marriage between autos and the oil and gas industry is bound for many more happy anniversaries.

PART IV

THE FUTURE

THE WORLD IN 2040

CHAPTER 15

FORECAST TROUBLES:

THE NUMBER KERFUFFLE

Time-machine back to 1981. Ronald Reagan succeeds Jimmy Carter as the fortieth president of the United States. MTV arrives in American living rooms for the first time, blasting rock ballad and thrash metal music videos to millions of teenagers sporting big hair and bigger shoulder pads. We see the first test-tube baby, we become aware of a disease called AIDS, and we watch as Major League Baseball teams go on strike crushing 38 percent of the schedule for fans.

It is also the year the price of oil begins its now-famous plummet, from $35 a barrel to less than $10 in just six short years.

The year before, the petroleum industry had invested $500 billion because it expected oil prices to rise 50 percent by 1985.[1] Unfortunately, this scenario is not unusual when it comes to forecasting. The fact is, the history of forecasting is dismal.

For example, according to Prakash Loungani with the International Monetary Fund, "The record of failure to predict recessions is virtually unblemished. Only two of the 60 recessions that occurred around the world during the 1990s were predicted a year in advance."[2] Likewise, a report called Consensus Forecasts noted that, out of 77 countries under consideration in 2008, economists did not predict recession for a single country in 2009.[3] And we all know how that turned out.

Likewise, after analyzing eight years of market calls by financial experts, a study by CXO Advisory Group found that the average guru was right about 47 percent of the time.[4] Turns out you would be better off letting a monkey make your stock picks—at least they might get you to 50 percent.

Forecasting in general is hard, and likewise, attempting to forecast the rate of EV adoption is no easy task.

Predicting the Unpredictable?

It should be no surprise, then, that forecasts dealing with the rate of EV adoption often disagree. A review of nine current scenarios on the future of EVs turned up some interesting contradictions. Figure 15.1 summarizes results from several publicly available reports from well-respected organizations ranging from those with aggressively pro-EV outlooks (e.g., Bloomberg) to more conservative viewpoints (e.g., Exxon).

Of the nine scenarios, predictions varied widely regarding how many EVs will be on the road in the next two decades. BP, being the most conservative, cites only 6 percent of vehicles on the road will be electric. By contrast, a study by International Monetary Fund (IMF) and Georgetown University indicated 93 *percent* of passenger vehicles would be EVs by midcentury. The majority of the reports fall somewhere in the middle, at somewhere in the range of 20–30 percent EV penetration by 2040.

	Percentage of EVs on Road by 2040*	Total Number of EVs by 2040*
BP (by 2035)	6%	100 million
BHP (by 2035)[A]	7%**	140 million
IEA, New Policies Scenario[B]	8%	> 150 million
EIA International Energy Outlook 2017 Low and High PEV Cases	8%–26%	160–520 million**
Exxon's 2017 Outlook for Energy***	~20%	400 million**
Electric Vehicle Outlook 2017 Bloomberg New Energy Finance[C]	34%	530 million
Carbon Tracker and the Grantham Institute at Imperial College London	35%	710 million**
IEA 450 Scenario[D]	>35%**	>710 million
IMF and Georgetown University (Method I)	93%	billion**

Figure 15.1. *Percentage/number of EVs on the road by 2040.*
Sources: [A] Clara Ferreira-Marques and Gavin Maguire, "BHP, World's Largest Miner, Says 2017 Is 'Tipping Point' for Electric Cars," Reuters, September 26, 2017, https://www.reuters. com/article/us-bhp-strategy/bhp-worlds-largest-miner-says-2017-is-tipping-point-for-electric-cars-idU.S.KCN1C10HO; [B]"World Energy Outlook 2016, Part B: Special Focus on Renewable Energy," International Energy Agency, 2016, https://www.iea.org/media/publications/weo/ WEO2016SpecialFocusonRenewableEnergy.pdf; [C]"Electric Vehicle Outlook 2017—Executive Summary," Bloomberg New Energy Finance, July 2017, https://data.bloomberglp.com/bnef/ sites/14/2017/07/BNEF_EVO_2017_ExecutiveSummary.pdf; [D]"World Energy Outlook 2016, Part B: Special Focus on Renewable Energy." Notes: *Unless otherwise noted; **Extrapolated based on assumption of 2 billion cars on road by 2040 (Berstein statistic);***With hybrids.

So what are we supposed to make of that? From 6 to 93 percent adoption? That's a ridiculous range.

Why so broad?

First, the methods and models that researchers used vary significantly from case to case. In the IMF/Georgetown study, EVs are considered a disruptive technology in much the same way coal was disruptive to wood, or cell phones were disruptive to landlines. The researchers argue, based on the rate by which automobiles overtook horses as a favored mode of transport, that EVs are poised to overtake ICEs much faster than almost anyone has predicted.

The IEA, on the other hand, provides its scenario based on policies and regulations—both in the pipeline and wished-for. Their most aggressive outlook, the IEA 450 Scenario, sets its EV targets by how many EVs need to be on the road to limit a global warming to no more than 2 degrees Celsius. (To keep global temperatures down in their scenario, a variety of CO_2 controlling measures must come to pass. One of these measures is the adoption of EVs.)

Other reports base their predictions on a variety of factors, including the availability of supporting infrastructure (charging stations); competition from biofuels, bikes, ride sharing, buses, more efficient ICEs, and mass transit; the ability of the electric grid to support the increased electricity demand; oil prices (as oil reaches peak demand, prices will fall, which bolsters demand); promises from car manufacturers to produce EVs; the increase in total future passenger vehicles; and government mandates and support. The latter can include incentives such as subsidies or penalties such as carbon taxes.[5]

One interesting market limitation is the supply of lithium, cobalt, and other rare earths, as well as the battery factories themselves. Auto manufacturers have announced targets for EV stock at between 9 and 20 million by 2020. But does the existing infrastructure allow for those targets to be met? According to the IEA Global EV Outlook 2017, "Attaining the mid-point of the estimated ranges for OEM announcements in 2025 would require the construction of roughly ten battery manufacturing facilities with the production capacity of the Tesla Gigafactory."[6]

That's quite a hurdle, but likely not insurmountable.

The Elusive Consumer Demand Prediction

Arguably the most important determining factor in the rate of EV adoption (and one that is conspicuously ignored in the IMF/Georgetown study) is consumer demand. So far, the demand for EVs has simply not materialized.

As we have seen many times over, when subsidies are removed, EV sales have tanked. When Denmark removed PEV subsidies in 2016, the country saw a 71 percent decrease in BEV sales and a 49 percent decrease in PHEV sales.[7] When subsidies in China were reduced, the leading EV manufacturer saw EV sales drop 34 percent in one quarter.[8] (This happens in the United States, too. Remember Georgia? Sales fell from 1,300 to 97 cars in just two months after the state scrapped its EV subsidies.)

Interestingly, most of the reports skirt around the demand issue, perhaps assuming it's a non-issue (we must clearly want them) or perhaps seeing consumer demand as too fickle to predict.

After all, forecasters have had no more luck predicting consumer demand than they have financial markets.

Consider the failure of high-dollar consulting reports to predict the pending boom of the budding cell phone market. If you were a potential cell phone investor in the early 1980s, you may have been greatly deterred by the report McKinsey produced for AT&T on the potential world cell phone market. Due to bulkiness of handsets, short-lived batteries, and other hard-to-overcome hurdles, the report predicted a market of only 900,000 cell phones by 2000. In 2000, 109 million cell phones had been sold.[9]

On the other hand, if you had invested in utilities based on 1970s electricity demand modeling, you could well have found yourself out a pretty penny. In 1974, U.S. electric utilities invested heavily in new generating plants, convinced of an upcoming 7 percent annual growth rate. In reality, the demand for energy grew at only 2 percent, leaving many projects cancelled or postponed, and electric companies hurting.[10]

Those who assume EV demand is a foregone conclusion seem to believe that a core driver of consumer purchasing decisions will be the growing fear of impending drought, fires, hurricanes, and earth-scorching heat waves. A 2015 Cone Communications/Ebiquity Global CSR Study gives this stance some hope. The study found that 81 percent of consumers stated they would make sacrifices to address social and environmental issues.[11]

But have consumers put their money where their mouth is? Time and time again, humans demonstrate that what we do and what we say we will do don't always line up. According to a Queensland University of Technology study, "Climate change beliefs have only a small

relationship with the extent to which people are willing to act in cli-
mate-friendly ways."[12]

That statement has certainly proven true for the carpool initiative.
Despite billions of dollars spent on mass transit, carpool lanes, and
driver education, Americans consistently choose driving alone over carpooling or mass transit. Seventy-five percent of Americans still drive to
work alone every day.[13]

The mismatch in belief and action is not only reflected in climate
change. The phenomenon holds true with other world-ending scenarios
as well. For example, in 1961, most Americans believed nuclear war was
likely. Yet only 200,000 families purchased fallout shelters (to put that
in perspective, the population at the time was around 180 million).[14] The
rest just hoped it wouldn't happen.

For EVs, this means that consumer demand must be driven by factors
other than a belief that EV ownership will help with climate change. Consumer demand must be spurred by more than a "green" badge of approval.

One potential candidate is cost savings. When EVs cost less than
ICEs, EV sales should increase. Although EVs are more expensive, it's
widely acknowledged that the battery pack costs are the bottleneck in
getting prices down. According to an IHS Markit study, battery packs are
expected to decline to a price point in the 2030s that will make EVs cost
competitive with ICEs. That's only ten years from now.[15]

Until then, many argue that due to the lower cost of EV maintenance,
we should already consider them price competitive with ICEs. To some
degree, this is true. However, customers in more developed countries
will be more likely to accept a higher ticket price, accepting that their
savings will come later over the lifespan of the vehicle.[16] In developing
nations, it may be difficult to justify a higher up-front cost, even if the
maintenance over time is significantly lower.

Of course, you don't necessarily have to be cheaper to pave the way
to buyers' hearts. Consumers will often purchase an item that is more
expensive, if it is "cool."

According to BP chief economist Spencer Dale, an EV "cool factor"
could spur sales to 450 million by 2035. That would take BP's estimate
from 6 percent of total passenger vehicles all the way to 25 percent.

Unfortunately, it's hard to know how cool consumers will find EVs. So far, sales results indicate "not very."

Even as Bloomberg writes headlines such as "Electric Car Sales Are Surging, IEA Reports," the reality is that EV sales are somewhat less impressive.[17] The IEA report Bloomberg references *does* say electric cars hit a milestone in 2016, reaching 2 million in global stock for the first time in history. What the editorial downplays is that that number represented an insipid 0.2 percent of total cars on the road at the time.[18] What Bloomberg failed to report entirely is that IEA reported 2016 to be a particularly bad year for sales: "2016 showed a slowdown in the market growth rate compared with previous years to 40 percent." *That would make 2016 the first year since 2010 that year-on-year electric car sales growth fell below 50 percent.*[19]

Commonsense Predictions

Statistics are a funny thing. You can make data say anything you want.

Coupled with the fact that the science of forecasting is difficult, when you add a layer of difficulty caused by human bias, you've really got a forecasting pickle.

Consider these two statements:

"The total number of EVs on the road will grow 4,900 percent by 2040."

"EVs will account for 6 percent of the global auto stock by 2040."

Can both be true? Of course. And yet one seems much more favorable to EV growth than the other. Combine statistic manipulation (hard for even the most neutral of journalists to avoid) and huge variations in predictions for EV adoption, and you can make a good case for an EV total revolution or an EV crash-and-burn scenario.

Which will it be? Only time will tell.

And *perhaps* time will bring about some commonsense forecasting. In his article entitled "Four Steps to Forecast Total Market Demand," Stanford University business professor William Barnett contends the reason for continued failure in forecasting is that forecasters often assume the past can be used to predict the future. He suggests a forward-looking approach, based on the following steps:[20]

1. Define the market.
2. Segregate demand into its core components.
3. Forecast the drivers of demand in each segment.
4. Conduct a sensitivity analysis to gauge the risk in your forecast.

Notice that all his steps are wrapped around a single concept of defining and predicting demand. It's interesting, when you consider how many other factors are being weighed in most of the current studies. Have we missed the most obvious, most driving, most determinant factor?[21] The factor that allows us to break seemingly impossible targets? The factor that truly unites us—more than any government action, rousing diatribe, or a deep-seated compulsion toward social responsibility? Have we forgotten to account for our desire to have something *better*?

Think VCRs, cell phones, and microwaves. These technologies all went from 10 to 80 percent U.S. market penetration in just ten years. Can EVs follow suit?

Certainly. Maybe.

We've already learned that, in 1900, electric cars made up one-third of the total automobile stock. Thomas Edison and Henry Ford were so enamored with the technology—quieter and safer than the early hand crank ICE models of the day—that they decided to work on a joint project. The result was to be an affordable EV, poised to push out the noisy and difficult-to-use early ICEs.

What happened? Something no one could have predicted. Mass production.

Well, technically mass production coupled with a tiff between the two now-famous inventors and helped by a few more hard-to-predict variables.[22] Purportedly Ford was determined only to use Edison's nickel-iron batteries in his vehicles, despite that the batteries' internal resistance was so high as to make them infeasible. When someone at the plant switched to lead-acid batteries without Ford's knowledge, he became irate and lost interest in the project.

Meanwhile, Ford had been working on several other projects and models, including the Model T Ford, which we all now know became the winning model. Coupled with the breakthrough mass production method, Ford was able to drive prices on the Model T to just $300 by 1925—compared to $850 just a few years earlier.[23] In addition, the invention of the electric starter made ICEs easier and safer to rev up in the morning. Finally, roads with gas stations began to pop up, and the American road trip became a cultural phenomenon—something not possible with electrics as few people had electricity outside of the city.

The EV's fate was sealed.

Our sleek, expensive, and silent transportation option silently disappeared—wiped so completely from our memories that most of us have no idea that the electric models of the early 1900s even existed.

Is the future of the EV more assured this time? Probably. Yet, despite efforts by the best statisticians and economists, trying to divine the future is still as tricky as it ever was.

Consider Stanford economists' somewhat off-the-wall scenario at RethinkX.[24] This group predicts 95 percent of all passenger miles will be driven by EVs by 2030 . . . far earlier than almost all other studies. How? Because these vehicles will also be autonomous, and hailing a robot-driven electric car at whim will be considerably cheaper than owning and driving your own automobile. (The average ICE vehicle costs $10,000 a year to own when driven just 15,000 miles.)

The autonomous EV fleet—favored to be an EV because EVs show large maintenance cost savings when driven more miles—will save the average American family $9,000 a year in road transportation costs. Not to mention the convenience of not having to drive. Imagine no more DWIs, distracted driving, or texting and driving. Commuters can use their commute time to work and can enjoy lower blood pressure as the stress of driving in traffic becomes a story for grandparents to tell their grandchildren.

That's something consumers can get behind. It's sexy. It's convenient. It's cheaper.

We plan for the EV to make inroads and displace ICEs, and it's highly likely that we will succeed—the pushes from governments, researchers, futuristic investors, and climate activists are all helping to drive prices lower. How fast they make inroads into our roads is anyone's guess. Price parity between ICEs and EVs will not be enough. Consumers will have to want these vehicles. They will need to want them because they are superior in some clear way—and not just environmentally.

That can happen.

If it does, electric cars will go the way of the iPhone. If it does not, the Blackberry might be a better analogy; the EV could still be beaten out by something exciting, new, and shiny—something we haven't even imagined yet.

CHAPTER 16

LET'S MAKE (UP) A DATE

If governments around the world have their way, by 2040 fossil fuel cars will be as extinct as the dinosaurs that gave their lives to eventually become oil and gasoline.

Under the banner of reducing their share of global CO_2 emissions, nations are uniting to put the brakes on the sale—and by extension, the manufacture—of gas-powered cars. As early as 2025, Norway intends to outlaw conventional cars, thanks to a head start in the form of support from its coalition government going back as far as 1990. Most countries have penciled in 2030 or 2040 on their calendars, including China, India, the Netherlands, France, and the United Kingdom. Cities, too, are in the race against the clock: Barcelona, Copenhagen, and Vancouver all plan to ban gas- and diesel-powered cars by 2030.

But even as more locales jump on the ICE-ban bandwagon, questions abound: How seriously can we take these tough goals and tight time-tables? Are these dates more fantasy than fact? And is it really the government's place to steer motorists to one type of vehicle over another? Isn't that a decision best left to consumers?

If converting the United States to EVs is a challenge, the barriers to converting the *world* is a Herculean task.

Sure, for the most part the ban has political will behind it today. But will future leaders still love it tomorrow? In two or three decades ahead, it's doubtful any of the politicians making decisions today will be in office; their legacies will have been consigned to history books. And some of their proposals—possibly including outright bans on conventional cars—are likely to be an asterisk in the pantheon of failed policies.

As Leslie Hayward, vice president of communications strategy and content for Securing America's Future Energy, said, "This could easily backfire. You can see consumers just totally rejecting this kind of thing. You're also forcing companies to deploy a technology on an imposed timeframe, rather than creating incentives for a technology to develop and then be able to compete appropriately because consumers want them, rather than being forced to buy them."[1]

Take a look at Ontario, Canada, for example, where change may be ahead—but it's not in the headlights. A December 2017 report by the CBC noted that lawmakers would like to see electric cars represent 5 percent of all passenger vehicles sold in the province by 2020. Then, it was less than 1 percent.

It'll never happen, experts say.

"The chances of meeting it aren't low, they're zero," the CBC quoted auto industry analyst Dennis DesRosiers. "In the auto sector all roads lead to electric. It just happens to be that the road to serious acceptance of them is probably at least 2030 and more likely 2040, 2050."[2]

That's despite $75 million in rebates to vehicle owners, tens of millions of dollars in subsidies, an installed network of charging stations, and an EV education center that cost $1 million.

India: Is Fifteen Years Realistic?

India is no Norway. The countries are polar opposites in more ways than where they sit on the globe.

Their climates are vastly different. They're dissimilar in terms of population density, pollution levels, economy, and culture. You might be able to make the case that India's naan is similar to Norway's lomper in that they're both flatbreads similar to the tortilla, but that's not a lot to draw on.

Norway is considered the world's electric car capital, with more than 121,000 EVs, hybrids, and even some hydrogen-cell cars plying the nation's roads. Thanks to exemptions on purchases and import taxes,[3] Norway even beat its uptake goal by 20 percent, expecting only 100,000 nonconventional cars in circulation by 2019. Growth shows no signs of stopping: March 2019 saw 11,518 EVs registered, nearly twice the number in March 2018, which was nearly twice as many as March 2017. The total share of all-electric cars grew to 31 percent in 2018.[4] As of February 2020, the country had reached 68 percent market share for plug-in passenger vehicles, with BEVs comprising just under half of the overall market, according to Clean Technica. ICEs continued to decline to just over 20 percent of market share.

Given this ostensible progress, one could reasonably expect that carbon dioxide emissions have fallen dramatically in the Scandinavian nation. Yet evidence suggests otherwise. Anders Skonhoft, an environmental economist at the Norwegian University of Science and Technology, told *The New York Times* that emissions were down by *no more than one-tenth of 1 percent* in 2015. And when you compare that to estimates that the total value of subsidies works out to about $13,500 a year per electric car over each vehicle's life, it doesn't look like Norway is getting much for its money.[5] Or shall we say, its taxpayers' money.

Not content to waste its own citizens' krone without full confidence that ICE vehicles are the way to go, Norway has offered to plug India— with 250 times the population—into its EV knowledge base. There are also rumors that Tata Power, India's largest electricity provider, is talking with a Norwegian company that builds charging infrastructure.

But is India ready for an EV revolution? Even with Nitin Gadkari,

India's minister of Road Transport and Highways, telling Indian auto-makers that he will bulldoze the initiative forward, the country is plagued by adoption barriers that make phasing out gas and diesel cars in less than 15 years an unlikely goal. The cards all just seem stacked against it.

First, 22 million conventional cars are sold in India each year. Next, the country's first EV charging station in Nagpur, a collaboration between local cab aggregator Ola and state-run Indian Oil Corporation, launched in November 2017—and takes as long as eight hours to fully home-charge a battery.[6] Last but certainly not least, hundreds of millions of the country's 1.32 billion people lack basic access to electricity.

World EV Adoption—Not Right Around the Corner

Converting the world to EVs isn't like pushing an "easy" button: There's no single, one-size-fits-all solution mostly because there's not just one identifiable problem. Circumstances and challenges differ from country to country. However, there are some common types of barriers, varying only in level of intensity. Among them are—

- Dependence on policy
- Lack of charging density
- Lack of electricity or grid infrastructure
- Challenges for certain types of vehicles
- Lack of familiarity
- Personal preferences
- Lobbyists
- Job loss
- Pushback from carmakers in oil-dependent economies

Let's measure how high each hurdle really is.

DEPENDENCE ON POLICY

In 2015, India introduced Faster Adoption and Manufacturing of (Hybrid &) Electric Vehicles (FAME) to promote clean fuel technology across all vehicle segments—cars, buses, light commercial trucks, two-wheelers, and so on. The program focused on technology development, demand creation, pilot projects, and charging infrastructure.[7]

Unfortunately, FAME essentially fizzled out. In two years, only about 2,000 of India's best-known brand of electric cars—the Mahindra Reva—were sold. Battery-operated two-wheelers didn't fare any better: In 2015 and 2016, there were only 20,000 sold compared to more than 15 million gasoline-powered models.[8]

In 2017, the government of Indian prime minister Narendra Modi decided to take a do-over. By December, it said, it would roll out a national policy that would set standards and specifications for the vehicles and provide guidelines for incentivizing their use. It would also include clarity on the rules regarding the manufacture and subsidies of batteries.

Policy-backed incentives are transitory by nature, which means their effects aren't lasting. For example, when Hong Kong stopped giving Tesla purchasers a tax break worth about $55,000, sales dropped to nothing. It took considerably less to get some Americans to break from buying EVs: Sales fell 80 percent when the state of Georgia repealed a $5,000 EV tax credit.[9] In 2016, when the government of the Netherlands reduced the amount of the exemptions for PHEVs in hopes it would spur sales of BEVs, the bottom dropped out of the market instead. During the first half of the year, business was off 73 percent.

On the flip side, we can look at Italy's latest endeavors. New regulations that took effect on March 1, 2019, offer subsidies for buying an EV up to €6,000—while introducing a new tax on ICE vehicles based on emissions. The more emissions, the higher the tax: €1,000 for cars that emit 161–175 grams of carbon dioxide per kilometer . . . all the way up to €2,000 for cars that emit 201–250 grams of carbon dioxide per kilometer.[10] And guess what? Almost immediately upon taking effect, EV sales in Italy skyrocketed to almost 1 percent of market share. April 2019 registrations of new plug-ins topped 1,600, a whopping increase of 271 percent over April 2018.[11]

This indicates that when free market governments set top-down tar-gets and incentivize adoption through subsidies (and what amounts to penalties for noncompliance), consumers aren't the ones creating product demand. Remove the carrot, and the stick just looks like something you get hit over the head with. People go right back to their old buying behaviors.

LACK OF CHARGING DENSITY

One reason acceptance is so high in Norway—besides massive subsi-dizing—is that the country has an advantage few others can claim: Of the Norwegians who drive an electric car, nearly all of them—some 96 percent[12]—have access to a charging station in their own home or apart-ment. For the rest, electricity is as close as the nation's well-established charging system, which includes the world's largest EV fast-charging sta-tion, located in a rural town some 60 kilometers north of Oslo.

Charging density isn't an issue in Norway, but for the rest of the world, the average driver can go more than twice as far in a conven-tional auto than an EV before they have to think about stopping to fuel up. While people driving conventional vehicles can venture far and wide knowing they'll easily find gas stations along the way, EV drivers must map out the location of charging stations and calculate charging time into their travel plans. It's no wonder that more people surveyed said the lack of charging stations would keep them from buying an electric car than any other reason, even price.

An MIT report by Rebecca Linke backed the public sentiment: "Even in more-developed countries, access to charging stations still places lim-its on EV adoption as does the length of time it takes to charge an electric battery."[13]

Consider France, for example. Since 2013, the government has been funding the installation of public charging stations in hopes of meeting its EV goals. But progress has been as slow as an escargot: The highest charging density across the country is only 0.1 charging points for every 1,000 registered vehicles.[14]

Latin American nations face similar struggles. Although there have been EVs roaming the roads of Colombia since 2012, the capital city of

Bogota didn't see its first public charging station until 2015. Brazil fares better but still has only 50 charging stations nationwide, most near the major cities of São Paulo and Rio de Janeiro.[15]

And what about those less-developed countries? Back in India, the state of Telanganga government is working to incentivize the use of EVs while also focusing on initiatives to develop the ecosystem—that is, charging stations—that allows them to run. Currently, there are only about one hundred charging stations across the nation, giving rise to serious feelings of restraint. As Anand Murali, writing for FactorDaily, said, "Owning an electric car in India feels like having a vehicle on a leash. Your driving distance is always limited by the charge your battery holds."[16]

To MIT Sloan professor David Keith, lack of charging density is a make-or-break deal—and the solution won't be ready tomorrow.

"We need charging of electric vehicles to be as easy as driving a gasoline vehicle today," he told MIT writer Rebecca Linke. "That means building the actual charging stations themselves, and then having an electricity grid that can support this additional demand for electricity. If we want fast charging, that means we need to provide a lot of electricity, often in locations where the grid may not necessarily be built for it today."[17]

LACK OF ELECTRICITY OR GRID INFRASTRUCTURE

Keith's comment brings us to our next consideration: Is there enough electricity to charge a world running on electric cars?

In most places, the answer is no.

Think about this: There are 240 million people living without electricity in India. In sub-Saharan Africa, the number is closer to 600 million.

In both India and Africa, even people with access to electricity often find the grid weak or unreliable—Africa's power shortages are so well known that one writer said South Africa has "turned rolling blackouts into a national art form."[18] According to *The New York Times*, all of sub-Saharan Africa's power generating capacity is less than South Korea's, and "a quarter of it is unproductive at any given moment because of the continent's aging infrastructure."[19]

With estimates suggesting that charging a fully electric worldwide fleet would lead to 10–20 percent more electricity demand, it's hard to imagine how emerging economies could generate enough power—and, especially, clean power—to support widespread expansion of EVs.

Yet into this low-energy scenario, Indian prime minister Narendra Modi has said he wants every one of the nation's 7.4 million vehicles in 2025 to be electric—not to mention the millions more two-wheelers, three-wheelers, e-rickshaws, and so on.[20] That will require not only the construction of a nationwide rapid recharging infrastructure—currently, the country's AC charging stations take as long as eight hours to fully power up a sedan—but also a tremendous growth in generating capacity.

It's a classic chicken-and-egg scenario: Will anyone buy an EV if there's no way to charge it? And without sales of EVs climbing, who is going to spring for the charging infrastructure?

And what of those South African grid disruptions, called "load shedding"? They occur when producers can't make enough electricity to meet demand and simply interrupt supply to certain areas. South African BMW, which is working to get a foothold in the EV market, says it has an app for that.

"Load shedding happens in periods throughout the day and an electric car can be charged during off-peak times, like when it is parked and overnight," spokeswoman Thando Pato said in an interview with Wedaeli Chibelushi for *Business Chief.* "With our ConnectedDrive App, a customer can constantly monitor the status of the car, even when they are not in it. So if the car is connected to a charger and there is load shedding and it stops charging, the ConnectedDrive App will be able to read this."[21]

CHALLENGES FOR CERTAIN TYPES OF VEHICLES

Most governments are focused on light-duty vehicles for electrification. But this shortsighted view omits perhaps the biggest emitters: heavy-duty trucks that burn a lot of fuel and produce much higher amounts of CO_2 and other pollutants. In the United States, medium- and heavy-duty trucks consume roughly 15 percent of the country's petroleum, and freight trucks spew out about 40 percent of all vehicle emissions.[22]

Some scientists argue that converting heavy-duty vehicles to electricity could certainly help America contribute to lowering the world's greenhouse gases. But the changeover is unlikely to happen, at least based on today's EV technology. For one thing, it would take about 30,000 pounds of batteries to move the average 80,000-pound weight of a class 8 commercial truck,[23] which clearly affects payload capacity in addition to being blatantly impractical. Add to that constraints on price and range, and it's easy to see why the long-haul segment will likely be the last to go electric.

LACK OF FAMILIARITY

How many EV commercials did you see during the last Super Bowl? How many EVs do you spot on the street? And—perhaps most important—do you think you'd know an EV if you saw one?

One of the biggest challenges EV manufacturers face is simple lack of recognition. According to a 2017 survey by consumer research firm Altman Vilandrie & Co., 60 percent of Americans said they were "unaware of electric cars."[24] Note, they didn't say they were "uninterested" in EVs. They weren't "opposed to" them. They didn't even know that EVs were an option. And this is unlikely to change until carmakers start to do a better job getting the word out.

Environmental group The Sierra Club teamed with Northeast States for Coordinated Air Use Management to commission a study about the amount of money advertisers spend promoting EVs and found the answer to be, well, not much. The survey discovered that outside of California, which has the most ZEV requirements in the United States, auto companies and auto dealers focused much less on EV advertising than on advertising for conventional cars and trucks.[25]

Take Ford as an example. The Sierra Club study found that in 2015, Ford advertised its gasoline-powered Focus in about 4,750 instances on cable and broadcast TV to national audiences. That was nearly 25 times more than it put its Focus Electric in front of a national TV audience.

PERSONAL PREFERENCES

If you like the roar of the engine or the smell of diesel, the experience of driving an EV might be a turnoff. The cars are so quiet, in fact, that in 2016 the U.S. National Highway Traffic Safety Administration ruled they must produce more noise when traveling at slow speeds to prevent accidents with people who are blind or visually impaired.[26]

Beyond those sensory differences—and despite their high prices—EVs also seem to lack the sort of prestige status that attracts buyers to, say, a Mercedes or Porsche. One of the best examples of this comes from China, where the government promotes the purchase of PEVs by offering some of the largest monetary incentives around. Still, consumers aren't taking the bait, preferring expensive gasoline-powered SUVs, instead. As an example, in May 2017, SUV sales in China were up 17 percent on a year-over-year basis.

This reaction may reflect an ironic consumer bias against vehicles that are economical to operate. RAND Europe reported on multiple studies from around the *world*—note not just the United States—indicating there's a social stigma against fuel-efficient vehicles, EVs included. In the United States, for instance, cars with good fuel economy were often considered "cheap" and less appealing to middle- and upper-class buyers.

And, true to the idea that beauty is in the eye of the beholder, some people just don't like the way EVs look. When researchers in Denmark asked people, including EV owners, to describe their "prejudices" against the cars, respondents said they were uglier than conventional cars, looking more like children's toys than grown-up transportation.[27]

LOBBYISTS

Without a working knowledge of where biofuels come from, it might be difficult to fathom how sugarcane could possibly stand in the way of Brazil's uptake of EVs. However, the Brazilian sugarcane industry association has come out roundly against EV adoption. That's because, for more than four decades, Brazil has been making ethanol from sugarcane, and a lot of it. Year after year, the country satisfies 25 percent of the world's ethanol demand; its sugarcane ethanol program is considered the most successful alternative fuel to date.[28] Switch off the need for ethanol and sugarcane growers, processors, and the Brazilian economy as a whole will suffer, UNICA, the industry group, contends.

Although Brazil has historically been the world's largest exporter of ethanol, it also keeps a considerable amount for its own use. No other nation uses as much biofuel for road transport as Brazil, and most of the cars manufactured in the country, regardless of maker, have built-in flex-fuel capability, meaning they can run on gasoline or ethanol.

However, with fuel price caps keeping gasoline and diesel costs low, many consumers have already been driven away from more expensive ethanol. The idea of losing more drivers to electricity is a bitter pill to Brazil's sugarcane farmers and processors.

JOB LOSS

It's not just Brazilian sugarcane workers who will lose much of their livelihood in a post–conventional car world. Assembly line automation already slashed the number of people it takes to manufacture a car; now, the increase in the number of automakers feeling compelled to add EVs to their lineup—which require fewer parts and less time to build—is having a chilling effect on autoworkers.

As for other types of automotive businesses, some may need to shift gears, so to speak, to prepare for increased EV usage. Electric cars have about 25 percent fewer parts than conventional autos. Companies that make parts the EVs don't use will need to diversify.

Paul Eichenberg, a Detroit-area automotive strategy consultant, predicted that three-quarters of the industry's top one hundred suppliers will fold by 2030 unless they start preparing for an EV revolution. "The years between 2020 and 2030 will be the decade of electrification, and if suppliers don't develop an e-mobility strategy and make investments now to operate in that decade, they risk becoming road kill on the global automotive highway," Eichenberg said shortly after releasing his white paper, "Electrification Disruption: How Not to Get Shocked, Jolted and Fried by the Coming Shift in Automotive Power Sources."[29]

Among the changes suppliers will need to make, Eichenberg said, is a different approach to staffing. Instead of seeking people with engineering skill sets, they'll need to focus on people with electrical and software expertise.

A study commissioned by the German Association of the Automotive Industry showed that banning internal combustion engines from 2030 would affect more than 600,000 jobs in Germany directly or indirectly, or 10 percent of the nation's workforce.[30]

In the United States, where both Ford and General Motors are amping up EV production, the powerful United Autoworkers (UAW) union is reviewing the risk to jobs and is ready to push back, if necessary.

"We've been doing our due diligence to find out how much [electrification] means to us," UAW vice president Jimmy Settles said in an interview with *Automotive News*. "We put them on notice early that we want to be part of the process."[31]

PUSHBACK FROM CARMAKERS IN OIL-DEPENDENT ECONOMIES

If low oil prices can destabilize places like Russia, Saudi Arabia, Nigeria, and Venezuela, whose economies depend on resource revenues, what happens if electricity overtakes oil as a transportation fuel? Even with the rosiest EV predictions, the need for oil doesn't go away; in fact the experts at consulting firm IHS suggest that by 2030 the world will actually need 40 million barrels per day of new crude production, most of it to meet demand from emerging countries.[32]

But what if Bloomberg New Energy Finance is right? They expect the rise in EVs to displace oil demand by 2 million barrels as early as the start of the 2020s and 8 million barrels by 2040.[33] That may seem like a drop in the bucket compared to the IHS prediction for increasing demand, but David Koranyi, the director of the Atlantic Council's Eurasian Energy Futures Initiative, said it could create the same kind of oversupply that led to the calamitous price drop of 2014.[34]

Russia, where oil and fuel products account for 54 percent of overall exports and 21.3 percent of GDP, isn't willing to take any chances: Key business leaders are giving the EV idea a big *nyet* in hopes, it would seem, of discouraging consumer interest. Igor Sechin, chief executive officer of state oil company Rosneft PJSC, has taken a hammer to the idea of EVs becoming popular in Russia, calling the Tesla overvalued and saying market expectations are significantly overestimated.

"Until the electric transport industry becomes as user-friendly and attractive for consumers as the cars with internal combustion engines, the prospects for electric vehicles remain largely uncertain," Sechin said.[35]

Statistics would indicate the oil boss is right: According to AutoStat data, 50 electric cars were sold in Russia within the first eight months of 2017, and that was a 35 percent year-on-year increase.[36]

As far as oil-rich Saudi Arabia is concerned, the kingdom has begun socking away money for a post-oil economy in a public investment fund that will eventually control $2 trillion in assets.[37]

Yet senior officials at oil giant Saudi Aramco say they worry more about the growth of ride-sharing apps such as Uber than about the rise of EVs. They predict EVs will account for only between 10 and 20 percent of the market by 2040.[38]

Yasser Mufti, vice president of corporate planning, told the *Financial Times* he thinks ride sharing, which could lead to a reduction in private car ownership, is a far more advanced trend than either electric or self-driving cars.[39]

Incidentally, Saudi Aramco has a 5 percent stake in Uber.[40]

What these barriers suggest is that the journey to an all-electric road fleet by some arbitrary, probably unattainable date won't be an easy ride—if the trip is even necessary at all. When the EPA set pen to paper, it determined that all forms of transportation in the United States—cars, trucks, planes, trains, and boats—account for about a quarter of the nation's CO_2 emissions, meaning cars are just a fraction of the total. In the European Union, cars account for 12 percent of CO_2 emissions. As Norway's example proves, the payback in terms of emission reduction isn't quite what anyone thought, promised, or pledged it might be.

So the real question is, are grand government mandates like banning fossil fuel cars anything but wasteful and inefficient? How can federal leaders justify creating demand for electricity that won't be easily met, using taxpayers' money to build infrastructure and pay for subsidies, and actually increasing greenhouse gases in the energy-intensive process of manufacturing electric cars—when in exchange, they've only made a dent in CO_2 emissions?

CHAPTER 17

THE FUTURE IS COMING

The future is here. If you need proof, just take a trip Down Under to the Australian city of Ipswich, Queensland.

Once a booming coal-mining town (it has been called "the birthplace of Queensland coalmining"[1]), Ipswich has been gradually moving away from coal since at least the 1980s. Although it hasn't abandoned coal completely, Ipswich is now home to a variety of industries and employers: In addition to its robust agriculture sector, Ipswich boasts a Royal Australian Air Force base, a burgeoning film industry, and a $72 million facility for General Electric, which relocated its headquarters to Ipswich in 2015—just to name a few.

As of 2017, the population of Ipswich was approximately 200,000. Due to rapid growth and the arrival of new industries, the population is expected to double by 2031. As the city grows, the Ipswich City Council is investing heavily in technology infrastructure, closed-circuit safety cameras, cultural events, and other amenities that, as council member

Paul Tully explained, are consistent with "urban areas and adjacent urban areas."

To generate publicity for the near completion of the city's newest charging station, Tully took a ride in a self-driving Tesla Model S. During his ride, the Tesla demonstrated its ability to read speed limit signs, automatically accelerate, and steer and brake.

"It was probably the smoothest ride I had in any vehicle, ever," he told *The Queensland Times*. "The only scary bit was coming up behind a stopped vehicle."

It may be a few years before self-driving cars are commonplace on the streets of Ipswich—or any modern city, for that matter. Perhaps, though, it's only a matter of time. After all, in the not-too-distant past, EVs seemed like something straight out of *The Jetsons*—as unrealistic as flying cars and intelligent robot maids.

Today, technologies like self-driving cars seem like sci-fi—but to people like Paul Tully, today's futuristic novelties are tomorrow's normal.

"A lot of people fear the future but we need to embrace it," Tully told *The Queensland Times* after taking a spin in the self-driving Tesla. "It's an exciting time."[2]

EVs Level Up: AVs and the Road to Autonomy

Perhaps the future of self-driving cars isn't as far away as we think: Although not exactly commonplace, self-driving cars are becoming a reality. It's slow going, though, and we are likely years away from driverless commutes and fully automated road trips. And yet, the National Highway Traffic Safety Administration has already developed a multilevel classification system for what it calls "autonomous" vehicles (AVs).

The system begins with Level 0, which comprises vehicles that are 100 percent human operated. From there, each level represents a higher level of automation, some more futuristic than others: Level 1, for example, includes features like adaptive cruise control.

Level 2 vehicles offer partial automation: The car's automated system handles some steering and acceleration/deceleration while the driver monitors the driving environment and handles other driving tasks. Examples of Level 2 AVs available today include the Tesla Autopilot, Volvo Pilot Assist, Mercedes-Benz Drive Pilot, and Cadillac Super Cruise. These vehicles manage speed and steering under certain conditions and can even match their speed to the traffic ahead of them and follow curves in the road, under certain conditions.

Things start to get real at Level 3, where automated systems can conduct driving tasks *and* monitor the driving environment, although the driver still must be ready to assume control of the vehicle whenever the automated system requests it.

At Level 4, the vehicle can drive itself and monitor road conditions without the need for a human driver at the ready. The catch? These vehicles are limited to specific conditions and environments. As an October 2017 article in *Car and Driver* explained, "In a . . . privately owned Level 4 car, the driver might manage all driving duties on surface streets then become a passenger as the car enters a highway."[3]

The goal—the vehicles that inhabit the sparkling, futuristic cityscapes being promised—is Level 5. These vehicles can handle 100 percent of the driving and road-monitoring tasks, in all driving conditions. At Level 5, drivers are little more than passengers: There's no need to watch the road or take the wheel, no reason to keep an eye on weather conditions or even a "scary bit" such as approaching a stopped vehicle. In a Level 5 AV, the driver is free to eat, talk on the phone, conduct business meetings, or read. In fact, they don't even need to be awake—commuters are free to catch a quick nap en route to the office. The car simply doesn't need them.

But for now, we'll need to keep our eyes open: "Today, right now, the highest level of autonomy available to us is Level 3—not full autonomy, or even high autonomy, no matter what marketing materials or other automotive publications say,"[4] wrote automotive journalist Justin Hughes.

So, yes, people *are* riding in AVs, but it's important to understand that what they're describing is not a Level 5—or anything even in that ballpark. At this point, even Level 3 AVs are a rarity for consumers. In fact, the first production automobile that could, accurately, be described as a Level 3 AV is Audi's 2018 A8 sedan.

But as an article from *Seeking Alpha* pointed out, the Audi's AV capabilities are limited: "For one it is only meant to be activated in relatively slow-moving traffic (up to 37 mph) and only on divided roads where a physical barrier separates vehicles driving in the opposite direction."

Nevertheless, this car still represents a breakthrough. "Audi's solution is the first of its kind to require *absolutely no monitoring* of the vehicle as long as the driver is available to take over when the vehicle asks him/her to," the article explains. "While the vehicle is driving itself, the customer is free to do anything else, for example even watch a movie on the on-board TV screen."[5]

Before AVs can progress to Level 5, the technology needs more testing, more fine-tuning, and a few more breakthroughs. In other words, it needs more time.

"Autonomous technology is where computing was in the '60s, meaning that the technology is nascent, it's not modular, and it is yet to be determined how the different parts will fit together,"[6] Shahin Farshchi, a partner at the venture capital firm Lux Capital, told *Wired*. Farshchi has invested in self-driving startup Zoox and sensor-builder Aeva.

Perfecting AV Tech with Pizzas and People

One of the most significant roadblocks on the path to Level 5 autonomy that AV manufacturers need to address is sensor technology. "Self-driving cars need at least three kinds to function—lidar (works like radar, but with light from a laser), which can see clearly in 3D; cameras, for color and detail; and radar, which can detect objects and their velocities at long distances," *Wired* reporter Aarian Marshall explained. "Lidar, in particular, doesn't come cheap: A setup for one car can cost $75,000. Then the vehicles need to take the info from those pricey sensors and fuse it together, extracting what they need to operate in the world and discarding what they don't."[7]

Another growing concern is cybersecurity. In the film *The Fate of the Furious* (the eighth installment of the long-running *Fast & Furious* series), an evil hacker portrayed by Charlize Theron gains control of an army of unattended cars and uses them to unleash chaos on New York City. Although this scenario sounds over the top, car and tech industry analysts say it's not that far-fetched.

But even if AVs are still in their early stages, they're accelerating closer toward the all-important Level 5. Not only are manufacturers furiously working on perfecting AV technology; municipalities are also collaborating with them to prepare for the time when AVs are more advanced and accessible. By the end of 2017, nearly 50 cities around the globe had completed or were in the midst of AV trials.[8] People in Ann Arbor, Michigan, for example, were getting their Domino's Pizza deliveries from a self-driving Ford Fusion hybrid during the summer of 2017.[9] (A human-driven car was always there to prevent mishaps.) Most people, by the way, thanked the vehicle for their pizza.

One of the most heartwarming pilot programs has been taking place in the United Kingdom, where a research and development consortium called Flourish has been using a mix of simulators and vehicles to learn how senior citizens behave in AVs. "This new technology could help older people to keep their independence and social connections beyond the age at which they would stop driving a conventional car,"[10] wrote Kevin O'Malley of *Prospect*.

Of course, not all the AVs being developed and tested today are EVs, but some would argue that pretty soon, the two technologies will be a match made in heaven. First, AV manufacturers will need help, like all carmakers, complying with government emissions requirements: EVs can do that. "Then there are engineering reasons—electric vehicles are easier for computers to drive," Greg Gardner wrote in the *Detroit Free Press*. That's because EVs have fewer moving parts: essentially a battery, an inverter, and the electric motor, explains Levi Tillemann-Dick, author of *The Great Race: The Global Quest for the Car of the Future*. "An internal combustion engine contains 2,000 tiny pieces that have to be kept lubricated and they break every once in a while," Tillemann-Dick said during an interview with Gardner.[11]

EVs will also help cut costs for one of AVs' most likely early users: ride hailing services. Tillemann-Dick estimates that companies like Uber would cut their cost per mile as much as 80 percent by providing electric, instead of ICE, self-driving shuttles, because electricity would cost about half the price of gasoline.

But Andrew Hawkins of The Verge argues that EVs will present their own challenges during the early days of AV usage. "For an electric vehicle to stay on the road for up to 20 hours of autonomous ride-hailing requires daily fast charging," Hawkins writes. "That can wear down a vehicle's battery, although both GM and Tesla insist their respective batteries hold up under the pressure of daily fast charging. But studies have shown that the charging time required for a battery-electric vehicle in an urban mobility setting, like ride sharing or car sharing, will significantly eat into utilization, especially if it has to use Level 2 charging."[12]

EVs and the Internet of Things

One thing all AVs likely will have in common is interconnectivity. In other words, they'll be part of the Internet of Things (IoT).

IoT encompasses any everyday appliance, device, and machine that's connected to the Internet, like smart thermostats and personal assistant devices. With this connectivity, devices can interact with people (think wearables, like fitness trackers), and the networked devices can interact with one another (like the Amazon Echo that will turn your lights off and on).

For future AVs, this kind of device-to-device communication will be essential. "If your vehicle doesn't know about a traffic jam along its route . . . it'll get stuck in gridlock. That's where connectivity comes in," wrote Roberto Baldwin of Engadget. "When self-driving cars hit the road, they'll not only be computing juggernauts but also sharing data with everything all the time."[13]

Along with the ability to avoid traffic jams and road hazards, these cars will maintain ideal spacing between each other, making traffic congestion less likely in the first place. They'll also prevent collisions by communicating their positions and checking in with each other and the smart, sensor-laden roads they're on. They'll also use their sensor-enabled communication capabilities to check for parking availability—another feature that will help alleviate road congestion.

In addition to chatting with parking spaces, these "smart" AVs will be able to communicate with businesses. They'll order and pay for food at drive-through restaurants. They'll find the nearest available EV charging stations and pay for their transactions. And they'll help with their own maintenance by recognizing failing components, ordering replacements, and driving themselves to the mechanic.

IoT technology has the potential to play an increasingly valuable role in bolstering the cool factor in EVs—and making the cars appealing to more consumers. A present-day example is Volkswagen's BUDD-e, an EV that lets drivers access connected home devices.

"You can switch on the air conditioner in your home from your car, so that it has the perfect temperature before you enter," *ELE Times* stated. "You can open your security gates for friends or relatives, even when you're away. Talk of luxury."[14]

In the future, EVs likely will be in constant communication with smart homes, smart cities, smart retailers—giving people the ability to do everything from shopping for jewelry with their car's technology to sending their autonomous EVs to pick up the kids from school.

The Future of Charging

Self-driving cars and "smart" parking spaces are fun to think about, to be sure. But to many people in the industry, one of the most exciting things about the next generation of EV technology is the idea that charging inconveniences may no longer be an issue. Imagine, instead of estimating hours between charging stations on road trips, drivers will effortlessly juice up their battery while they zoom along the highway. Pads beneath the road's surface will top off battery charges as cars move over them. This "dynamic inductive charging" technology wouldn't work with fully drained batteries, but the potential still is huge.

"Topping up cars as they roll over the charging segments could mean exiting a highway with more power than you started with," Antuan Goodwin of CNET wrote. "However, the tech also has the potential to shape the way future EVs are built and how much they might cost. Widespread inductive charging for parking and for roads could mean that EVs might get away with using smaller, lighter batteries, since they'd be re-juicing more often, or a 40- to 50-mile stated range could be stretched significantly further while the road is supplying power along the way."

Qualcomm, a multinational semiconductor and telecommunications company headquartered in San Diego, installed dynamic charging technology in a 100-meter segment of test track in France in May 2017. During their demonstration, two Renault Kangoos showed how cars in the future will charge on the go. "To me, the biggest advantage is allowing cars with smaller batteries to make useful trips," Bill Von Novak, a member of the Qualcomm team that invented wireless charging, told Goodwin. "That includes picking up the kids on a busy day and getting a little bit of charge from the roadway or taking a small-battery EV from San Diego to LA on I-5 while the highway charges the car along the way. The main advantage of this is that it allows you to have ranges and do things with electric cars that otherwise you just couldn't do without spending a huge amount of money on batteries."[15]

And Qualcomm isn't the only company working on wireless EV charging. In 2007, a team of physicists from the Massachusetts Institute of Technology launched WiTricity to commercialize the wireless charging technology they developed for EV owners to charge their cars while parked. Since 2018, the company has worked with carmakers and licensed Tier 1 suppliers to bring this technology to market. WiTricity CEO Alex Gruzen said, "With no driver, who will plug in the vehicle to recharge it? The answer is clear: No plugs, no wires. Park-and-charge wirelessly and autonomously . . . with WiTricity technology."[16]

Wireless, in-road charging is already at work for public transportation routes in South Korea, where a 7.5-mile stretch of road has been charging bus batteries since 2013. The technology was developed by South Korea's Advanced Institute of Science and Technology.

Specially designed electric buses are operating on the road with bat-teries a third of the size of those usually used in EVs. Now the Israeli government is working with startup ElectRoad to establish a public bus route that uses an in-road wireless charging system.[17]

But the arrival of new highway systems that provide large-scale EV charging may take a bit longer simply because of the high costs involved. And charging pad providers will have details to work out, too, before governments start building roads with them. How, for instance, will the charging pads be able to charge multiple vehicles, with multiple designs, moving at varying speeds? What will happen when the ground is covered with snow and ice? And how will the charge providers get paid? (Stay tuned for a possible answer to the last question.)

But in-road EV systems may not be the only cool new charging tech-nology in drivers' futures. Before long, the Airbnb business model that has people opening their homes to strangers for lodging may surface in a similar incarnation: peer-to-peer charging services.

"Let's say a participating EV driver is looking to park and charge while picking up a poké bowl for lunch, and all of the public chargers in downtown Mountain View are taken," wrote Julian Spector of Greentech Media. "This driver could tap into the network, see an open charger in a driveway a few blocks away, and plug in there."[18]

Peer-to-peer charging already is a reality in Germany, and a pilot is in the works in California. The system relies on blockchain—the dig-ital ledger system that made Bitcoin possible—to verify transactions. "Blockchains are decentralizing accounting systems, where transac-tions are recorded across multiple computers,"[19] Ben Schiller of *Fast Company* explained.

Since blockchain was invented, the technology has been adapted for tracking digital units other than cryptocurrency, including electrons. Today, blockchain can record energy flows between charging units and cars, and in the future it might even be used to record flows between vehicles and in-road charging systems.

When Cars Fly: Predicting the Future

So, will EVs eventually be part of what now seems like a futuristic transportation ecosystem? Will self-driving EVs, flying cars, and a hyperloop become the new normal?

Well, that depends on who you ask.

Take flying cars. "The technology required for flying cars is finally here. The biggest challenge—creating safe and effective vertical take-off and landing (VTOL) vehicles (say, combining helicopter and plane)—is being solved with innovation in aircraft design and recent progress in autonomous flight and electric motors and batteries," wrote Valery Komissarov of TechCrunch. "We'll see flying cars in the mass market in around 10–15 years—this time is likely required to solve the problem of integration into public airspace and, in some sense, convince customers that the idea is not that much more crazy than, for example, ICOs (initial coin offerings/cryptocurrency) or VR games."[20]

It's safe to say that the world will need at least that much time to create, perfect, and accept transportation technologies of the future.

It's also safe to say that however that future shapes out, EVs have significant value to offer. Like *ELE Times* says, "From the first vacuum tube car radio, to Internet of Things, the revolutions in electronics have given the car industry an impetus toward countless possibilities."[21]

CONCLUSION

Above the displays of ancient Egyptian artifacts and elaborate Fabergé eggs, over a pendulum in perpetual motion and the idled skeletons of prehistoric beasts, there is a sort of petro-wonderland on the top floor of the Houston Museum of Natural Science.

At just about 30,000 square feet—almost the size of a football field—the Wiess Energy Hall is an interactive homage to all things oil and gas. What could be more appropriate in the city that bills itself the energy capital of the world?

Here, the facts about petroleum exploration and production have been given the kind of fantastic treatment that would make a Disney Imagineer proud. A ragtag bunch of robots populates a drilling rig. Visitors are virtually shrunk to the size of a grain of proppant sand, then transported into a hydraulic shale fracture. Kids—and yes, adults, too—race cleaning tools called pigs through a plastic pipeline and watch Plinko ball-sized oil molecules separate into sweet and heavy crude. Cheery, oil-and-gas-themed bubblegum pop emanates from the "Energy Jukebox."[1]

It's all good, animated fun, a glimpse into the processes and products that fuel one of the state's chief economic engines.

Two flights down, though, the mood is considerably more somber.

In the wildlife exhibits, the emphasis is on rare, endangered, and extinct species. Children who just moments earlier emerged from a manic, Day-Glo oil and gas wonderland stare in quiet reverence at drawings and grainy, centuries-old photographs of the passenger pigeon, the dodo, and the Tasmanian tiger, animals they've never seen in real life, and—due to a combination of evolution, predation, and human involvement—never will.

It can be difficult, even for adults, to reconcile the messages on these seemingly disparate museum floors. Oil is essential to our everyday lives. In a lot of ways, it really does make the world go 'round, just like the song says. At the same time, many of the world's largest oil and natural gas reserves are hidden beneath diverse and fragile ecosystems, making energy exploration and production there potentially harmful to animals and plants. Fortunately, an enlightened industry is working responsibly to prevent the kinds of ecological impact its detractors worry about, avoiding sensitive areas, using tested technologies to access hard-to-reach deposits, and minimizing impacts on water, land, air, and ice.

So it doesn't have to be an either-or proposition. We can keep the world moving forward while at the same time preserving natural resources for our children—and our children's children.

There's room for innovation and compromise, for each of us to help the world on our own terms. We can have a new and improved energy mix that still includes fossil fuels and that also creates a cleaner energy future. Although we can't go back in time to save the Tasmanian tiger or the passenger pigeon, we can work to make sure that if anything else faces the same fate, humans haven't had a hand in it.

Leveraging the gee-whiz science of electric cars is one step in that direction. But, like all massive technology changes before it, we will likely encounter our share of failures before we find the right formula for success. Remember, even the transition from horses to horsepower meant stepping in plenty of manure first.

Some of these challenges will be financial, others technical or social. And, yes, there will be political issues to contend with.

But it'll be worth it.

Imagine a future where fully autonomous EVs are commonplace and the electricity that powers them comes from clean natural gas, wind, or solar—with coal long since relegated to the dustbin, even in Africa. The need for hydrocarbons hasn't gone away—remember all those petroleum-based car parts, tires, and roads—but electrification, automation, and digitalization have transformed the way oil and gas are safely located, extracted—even from the ocean depths—and delivered to users. Deploying smart machines in place of humans has reduced risk, increased efficiency, and made production cleaner and more economical than ever. Advanced analytics are generating best practices, literally from beneath-the-ground up.

A hands-free trip with friends into the city leaves you free to blast the virtual aliens surrounding the car courtesy of a new video game down-loaded to altered reality (AR) glasses. A virtual "bad guy" leaps onto the hood. *Pew!* A direct hit! As the downed alien disappears in a mirage of pixels, three more take his place. A call for backup encourages a passenger to join the fight via his own AR device. *Pew! Pew!*

Meanwhile, in the back seat, another passenger takes a moment to check an online order she placed earlier that day—the delivery drone is already on its way. In fact, if you weren't so busy with your game, you might have noticed the sky is full of drones, branded with company logos and loaded with shipments of documents, merchandise, and supplies.

Finally, the car rolls to a stop in front of the natural history museum. You follow the crowd in and up to the second floor. The excitement is palpable: Today is the grand opening of the exhibit about animals brought back from the brink of extinction. Green technology has played a part by reducing emissions, but so have individuals by slashing their own carbon footprints. Children jockey for places in front of holographs of once-imperiled animals like king penguins,[2] polar bears, and snow leopards.

Two flights up, the energy exhibit is still going strong, but now the centerpiece is a model of an automated oil field that runs nearly self-sufficiently. With operations controlled from a distance, helicopters that used to shuttle crews no longer hover over platforms but hang overhead on display instead. Models show how underwater drones and unmanned submersibles monitor integrity and inspect equipment, sending video feeds back to home base.

And on the third floor in between, there's a metaphorical and literal link between the displays above and below: an exhibit about the modern electric car.

NOTES

INTRODUCTION

1. "The Morrison Electric Automobile & The William Morrison Co.," American Automobiles, n.d., www.american-automobiles.com/Electric-Cars/Morrison-Electric.html

2. Rebecca Matulka, "The History of the Electric Car," US Department of Energy, September 15, 2014, https://www.energy.gov/articles/history-electric-car

3. "Timeline: History of the Electric Car," Week of October 30, 2009, PBS, http://www.pbs.org/now/shows/223/electric-car-timeline.html

4. "Fast-Forwarding to a Future of On-Demand Urban Air Transportation," Uber, October 27, 2017, https://www.uber.com/elevate.pdf

5. "2018 Outlook for Energy: A View to 2040," ExxonMobil, 2018, http://cdn.exxonmobil.com/~/media/global/files/outlook-for-energy/2018/2018-outlook-for-energy.pdf

6. Grant Smith, "U.S. to Dominate Oil Markets after Biggest Boom in World History," Bloomberg, November 13, 2017,

https://www.bloomberg.com/news/articles/2017-11-14/
iea-sees-u-s-shale-surge-as-biggest-oil-and-gas-boom-in-history

7. Lizzie Wade, "Tesla's Electric Cars Aren't as Green as You Might Think," *Wired*, March 31, 2016, https://wired.com/2016/03/teslas-electric-cars-might-not-green-think

8. "Subsidizing Electric Vehicles Inefficient Way to Reduce CO_2 Emissions: Study," phys.org, June 22, 2017, https://phys.org/news/2017-06-electric-vehicles-inefficient-co2-emissions.html

9. Concordia University, "Study Reveals that Green Incentives Could Actually be Increasing CO_2 Emissions," phys.org, June 7, 2017, https://phys.org/news/2017-06-reveals-green-incentives-co2-emissions.html

10. Magdalena Dugdale, "European Countries Banning Fossil Fuel Cars and Switching to Electric," Road Traffic Technology, August 1, 2018, https://www.roadtraffic-technology.com/features/european-countries-banning-fossil-fuel-cars/

11. Michael J. Coren, "Nine Countries Say They'll Ban Internal Combustion Engines. So Far, It's Just Words," Quartz, August 7, 2018, https://qz.com/1341155/nine-countries-say-they-will-ban-internal-combustion-engines-none-have-a-law-to-do-so/

12. Jules Verne, *Journey to the Center of the Earth*, Ch. XXXI: Preparations for a Voyage of Discovery, 1864.

PART I: THE DEBATE

CHAPTER 1: CLIMATE CHANGE WARS: EV VS O&G

1. Justin Worland, "Climate Change Used to Be a Bipartisan Issue. Here's What Changed," *Time*, July 27, 2017, http://time.com/4874888/climate-change-politics-history/

2. Cary Funk and Brian Kennedy, "For Earth Day, How Americans See Climate Change in 5 Charts," Pew Research Center, April 19, 2019, https://www.pewresearch.org/fact-tank/2019/04/19/how-americans-see-climate-change-in-5-charts/

3. Mark P. Mills, James B. Meigs, and John Stossel, "The Green New Deal's Bad Science," *City Journal*, April 22, 2019, https://www.city-journal.org/stossel/green-new-deal

4. Gary Sernovitz, "Can Liberals Frack?" *The New York Times*, April 11, 2016, https://www.nytimes.com/2016/04/11/opinion/can-liberals-frack.html

5. Amanda Scott, "President Obama Talks Energy at the State of the

Union 2013," U.S. Department of Energy, February 13, 2013, https://energy.gov/articles/president-obama-talks-energy-state-union-2013

6. Mark Watts, "Streets Can Kill Cities: On the Fossil-Free-Fuel Streets Declaration," CityMetric, December 18, 2017, https://www.citymetric.com/transport/streets-can-kill-cities-fossil-free-fuel-streets-declaration-3555

7. Tom Johnson, "In NJ, Electric Vehicles Could Be Key to Economic, Environmental Progress," NJ Spotlight, December 18, 2018, http://www.njspotlight.com/stories/17/12/17/in-nj-electric-vehicles-could-be-key-to-economic-environmental-progress/

8. Joshua D. Rhodes and Michael E. Webber, "Commentary: The Solution to America's Energy Waste Problem," *Fortune*, December 18, 2018, http://fortune.com/2017/12/18/electrification-energy-u-s-economy/?xid=gn_editorspicks

9. G.E. Miller, "Electric Vehicle Tax Credits: How Much, How to Claim, & Manufacturer Phaseouts (Already Happening!)," 20Something Finance, April 21, 2019, https://20somethingfinance.com/electric-vehicle-tax-credit/

10. Frank Watson, "December Global Electric Vehicle Sales Set New Record," S&P Global Platts, February 11, 2019, https://www.spglobal.com/platts/en/market-insights/latest-news/electric-power/021119-december-global-electric-vehicle-sales-set-new-record-sampp-global-platts-data

11. Jose Pontes, "Global Top 20 - December 2018," EVSales.com, January 31, 2019, http://ev-sales.blogspot.com/2019/01/global-top-20-december-2018.html

12. Roland Irle, "Global Plug-in Sales for 2017-Q4 and the Full Year," EVvolumes.com, January 2018, http://www.ev-volumes.com/country/total-world-plug-in-vehicle-volumes/

13. Angela Jameson, "Darker Side of Electric Cars in Spotlight," *The National*, November 19, 2017, https://www.thenational.ae/business/technology/darker-side-of-electric-cars-in-spotlight-1.676999

14. Rakteem Katakey, "Shell Takes Exxon's Cash-Flow Crown as Earnings Beat Estimates," Bloomberg Markets, November 2, 2017, https://www.bloomberg.com/news/articles/2017-11-02/shell-takes-exxon-s-cash-flow-crown-as-earnings-beat-estimates

15. "Royal Dutch Shell capital expenditure from 2011 to 2018," Statista, 2019, https://www.statista.com/statistics/561546/royal-dutch-shell-capital-expenditure/

16. Jonathan Haidt, *The Happiness Hypothesis: Finding Modern Truth in Ancient Wisdom*, New York: Basic Books, 2006.

17. Alanna Petroff, "Volkswagen Used to Love Diesel. Not Anymore," CNN Money, December 11, 2017, http://money.cnn.com/2017/12/11/investing/volkswagen-vw-diesel-subsidies-ceo-matthias-muller/index.html

18. Alessandra Potenza, "Excess Pollution from Diesel Cars Leads to 5,000 Premature Deaths a Year in Europe," The Verge, September 18, 2017, https://www.theverge.com/2017/9/18/16328092/diesel-cars-air-pollution-premature-deaths-europe-dieselgate

19. Melissa Eddy and Jack Ewing, "As Europe Sours on Diesel, Germany Fights to Save It," *The New York Times*, August 2, 2017, https://www.nytimes.com/2017/08/02/business/energy-environment/germany-diesel-car-emissions.html

20. Damian Carrington, "Diesel Cars Emit 10 Times More Toxic Pollution Than Trucks and Buses, Data Shows," *The Guardian*, January 6, 2017, https://www.theguardian.com/environment/2017/jan/06/diesel-cars-are-10-times-more-toxic-than-trucks-and-buses-data-shows

21. Melissa Eddy and Jack Ewing, "As Europe Sours on Diesel, Germany Fights to Save It."

CHAPTER 2: CARBON À LA CARTE: THE TRANSPORTATION SMORGASBORD

1. David L. Barnes and Arnold R. Miller, "Fuelcell-Powered Front-End Loader Mining Vehicle," DOE Hydrogen Program FY 2004 Progress Report, 2004, https://www.hydrogen.energy.gov/pdfs/progress04/ve3_barnes.pdf

2. Xinhua, "Hydrogen Fuel Cells Power Tangshan Tram," *Global Times*, October 27, 2017, http://www.globaltimes.cn/content/1072303.shtml

3. "Germany Launches World's First Hydrogen-Powered Train," *The Guardian*, September 17, 2018, https://www.theguardian.com/environment/2018/sep/17/germany-launches-worlds-first-hydrogen-powered-train

4. "The Longest Classic Cars of All Time," Classic Car Labs, May, 31, 2016, http://classiccarlabs.com/2016/05/31/longest-american-classic-cars/

5. "Fact Sheet: Driving to 54.5 MPG: The History of Fuel Economy," The Pew Charitable Trusts, April 20, 2011, http://www.pewtrusts.org/en/research-and-analysis/fact-sheets/2011/04/20/driving-to-545-mpg-the-history-of-fuel-economy

6. "United States: Light-Duty Vehicles: GHG Emissions & Fuel Economy,"
    DieselNet, n.d., https://www.dieselnet.com/standards/us/fe_ghg.php

7. "Regulations for Emissions from Vehicles and Engines,"
    U.S. Environmental Protection Agency, n.d., https://www.
    epa.gov/regulations-emissions-vehicles-and-engines/
    regulations-greenhouse-gas-emissions-passenger-cars-and

8. Chris Edwards, "Privatizing Amtrak," Downsizing the Federal
    Government, October 17, 2016, https://www.downsizinggovernment.
    org/transportation/privatizing-amtrak

9. Maggie Astor and Allya Yourish, "Amtrak Faces a High-Profile
    Derailment. Again," *The New York Times*, December 18, 2017, https://
    www.nytimes.com/2017/12/18/us/amtrak-derailments.html

10. National Railroad Passenger Corporation and Subsidiaries (Amtrak),
    "Management's Discussion and Analysis of Financial Condition and
    Results of Operations and Consolidated Financial Statements with
    Report of Independent Auditors," Fiscal Year 2018, https://www.amtrak.
    com/content/dam/projects/dotcom/english/public/documents/
    corporate/financial/Amtrak-Management-Discussion-Analysis-Audited-
    Financial-Statements-FY18.pdf

11. Ralph Buehler, "9 Reasons the U.S. Ended Up So Much More Car-
    Dependent Than Europe," CityLab, February 4, 2014, https://www.
    citylab.com/transportation/2014/02/9-reasons-us-ended-so-much-
    more-car-dependent-europe/8226/

12. "Passenger cars in the EU," Eurostat, April 2018, https://ec.europa.eu/
    eurostat/statistics-explained/index.php/Passenger_cars_in_the_EU

13. "Passenger Transport Statistics," Eurostat, April 2017, http://
    ec.europa.eu/eurostat/statistics-explained/index.php/
    Passenger_transport_statistics

14. Yonah Freemark, "Have U.S. Light Rail Systems
    Been Worth the Investment?" CityLab, April 10, 2014,
    https://www.citylab.com/transportation/2014/04/
    have-us-light-rail-systems-been-worth-investment/8838/

15. "Ride Sharing," Statista, n.d., https://www.statista.com/
    outlook/368/109/ride-sharing/united-states#

16. "Regional Distribution of Rideshare Apps," Techcabal, November 2016,
    http://techcabal.com/wp-content/uploads/2016/11/Ridesharing-apps-
    Iinfographics.jpg

17. "Latest Telecommuting Statistics," GlobalWorkplaceAnalytics.com, June
    2017, http://globalworkplaceanalytics.com/telecommuting-statistics

18. Alex Taylor III, "10 Alternatives to the Gasoline-Powered Engine," *Fortune*, November 1, 2013, http://fortune.com/2013/11/01/10-alternatives-to-the-gasoline-powered-engine/

19. "Hybrid and Plug-In Electric Vehicles," n.d., U.S. Department of Energy Alternative Fuels Data Center, https://www.afdc.energy.gov/vehicles/electric.html

20. "Financing Models: Propane Autogas Vehicles and Infrastructure," North Carolina Clean Energy Technology Center at N.C. State University, n.d. https://nccleantech.ncsu.edu/wp-content/uploads/Propane-Finance-Models.pdf

21. Alex Taylor III, "10 Alternatives to the Gasoline-Powered Engine."

22. Stuart Nathan, "Electric, Hybrid or Alternative Fuel Cars— Which Will Prevail?" October 3, 2013, The Engineer, https://www.theengineer.co.uk/issues/october-2013-online/electric-hybrid-or-alternative-fuel-cars-which-will-prevail/

CHAPTER 3: ENERGY IN/ENERGY OUT: EV'S CARBON FOOTPRINT

1. Jon Grinspan, "The Saloon: America's Forgotten Democratic Institution," *The New York Times*, November 26, 2016, https://www.nytimes.com/2016/11/26/opinion/sunday/the-saloon-americas-forgotten-democratic-institution.html

2. David Morris, "Solar Briefly Topped 50% of California Electricity in March, Driving Rates Below Zero," *Fortune*, April 8, 2017, http://fortune.com/2017/04/08/solar-california-electricity/

3. "Total System Electric Generation," California Energy Commission, June 21, 2018, http://www.energy.ca.gov/almanac/electricity_data/total_system_power.html

4. Damian Carrington, "China's Coal-Burning in Significant Decline, Figures Show," *The Guardian*, January 19, 2016, https://www.theguardian.com/environment/2016/jan/19/chinas-coal-burning-in-significant-decline-figures-show

5. Chris Buckley, "China Burns Much More Coal Than Reported, Complicating Climate Talks," CNBC, November 3, 2015, https://www.cnbc.com/2015/11/03/china-burns-much-more-coal-than-reported-complicating-climate-talks.html

6. Maria Gallucci, "Coal-Fired Electricity Is at Its Lowest Since Officials Started Keeping Track," Mashable, February 7, 2017, https://mashable.com/2017/02/08/america-coal-carbon-emissions-data/#y1ewuLqd65qq

7. Gayathri Vaidyanathan, "How Bad of a Greenhouse Gas Is Methane?"

Scientific American, December 22, 2015, https://www.scientificamerican.com/article/how-bad-of-a-greenhouse-gas-is-methane/

8. John Schwartz, "Gas Utilities Reduce Leaks of Methane, Study Finds," *The New York Times*, March 31, 2015, https://www.nytimes.com/2015/04/01/science/earth/gas-utilities-reduce-leaks-of-methane-study-finds.html

9. Ramón A. Alvarez, Stephen W. Pacala, James J. Winebrake, William L. Chameides, and Steven P. Hamburg, "Greater Focus Needed on Methane Leakage from Natural Gas Infrastructure," Proceedings of the National Academy of Sciences, April 2012, http://www.pnas.org/content/109/17/6435

10. Anrica Deb, "Why Electric Cars Are Only as Clean as Their Power Supply," *The Guardian*, December 8, 2016, https://www.theguardian.com/environment/2016/dec/08/electric-car-emissions-climate-change

11. "Emissions from Hybrid and Plug-In Electric Vehicles," U.S. Department of Energy Alternative Fuels Data Center, n.d., https://www.afdc.energy.gov/vehicles/electric_emissions.php

12. Steve Hanley, "10 Myths About Electric Cars. Are Any of Them True?" Gas2, October 16, 2017, https://gas2.org/2017/10/16/10-myths-about-electric-cars-any-true/

13. "Swedish Study Calls for Smaller EV Batteries, Finds Tesla More Polluting than an 8-Year-Old Car," Autovista Group, June 16, 2017, https://www.autovistagroup.com/news-and-insights/swedish-study-calls-smaller-ev-batteries-finds-tesla-more-polluting-8-year-old

14. Sean Szymkowski, "Bizarre Swedish Study Claims Electric Cars Are Worse for the Environment," Green Car Reports, June 28, 2017, https://www.greencarreports.com/news/1111266_bizarre-swedish-study-claims-electric-cars-are-worse-for-the-environment

15. Anthony Watts, "Tesla Car Battery Production Releases as Much CO_2 as 8 Years of Gasoline Driving," Watts Up with That, June 20, 2017, https://wattsupwiththat.com/2017/06/20/tesla-car-battery-production-releases-as-much-co2-as-8-years-of-gasoline-driving/

16. Ezra Dyer, "That Tesla Battery Emissions Study Making the Rounds? It's Bunk," *Popular Mechanics*, June 22, 2017, http://www.popularmechanics.com/cars/hybrid-electric/news/a27039/tesla-battery-emissions-study-fake-news/

17. Frequently Asked Questions: How Much Carbon Dioxide Is Produced from Burning Gasoline and Diesel Fuel?" U.S. Energy Information Administration, May 19, 2017, https://www.eia.gov/tools/faqs/faq.php?id=307&t=11

18. "Properties of Fuels," U.S. Department of Energy Alternative Fuels Data Center, n.d., https://afdc.energy.gov/fuels/fuel_comparison_chart.pdf

19. Lizzie Wade, "Tesla's Electric Cars Aren't as Green as You Might Think," *Wired*, March 31, 2016, https://www.wired.com/2016/03/teslas-electric-cars-might-not-green-think/

20. Norman Mayersohn, "The Internal Combustion Engine Is Not Dead Yet," *The New York Times*, August 17, 2017, https://www.nytimes.com/2017/08/17/automobiles/wheels/internal-combustion-engine.html?_r=0

21. Chris Bruce, "Mazda3 With Skyactiv-X Engine Not Coming To U.S. Anytime Soon," Motor1.com, April 24, 2019, https://www.motor1.com/news/346378/mazda-skyactiv-x-engine-delay

22. David Biello, "Electric Cars Are Not Necessarily Clean," *Scientific American*, May 11, 2016, https://www.scientificamerican.com/article/electric-cars-are-not-necessarily-clean/

23. Steve Hanley, "10 Myths About Electric Cars. Are Any of Them True?"

24. David Noland, "Does the Tesla Model S Electric Car Pollute More Than an SUV?" *Popular Science*, May 31, 2013, https://www.popsci.com/cars/article/2013-05/does-tesla-model-s-electric-car-pollute-more-suv#page-3

25. Askar Sheibani, "Rare Earth Metals: Tech Manufacturers Must Think Again, and so Must Users," *The Guardian*, March 26, 2014, https://www.theguardian.com/sustainable-business/rare-earth-metals-upgrade-recycle-ethical-china

26. "If All U.S. Cars Suddenly Became Electric, How Much More Electricity Would We Need?" (question posed and answered on Quora), *Slate*, May 2, 2014, http://www.slate.com/blogs/quora/2014/05/02/electric_vehicles_how_much_energy_would_we_need_to_fuel_them.html

CHAPTER 4: IT'S COMPLICATED: CLIMATE CHANGE, CO₂, COST

1. "Manatee Reclassified from Endangered to Threatened as Habitat Improves and Population Expands—Existing Federal Protections Remain in Place," U.S. Fish & Wildlife Service, March 30, 2017, https://www.fws.gov/news/ShowNews.cfm?ref=manatee-reclassified-from-endangered-to-threatened-as-habitat-improves-a&_ID=36003

2. Lucy Schouten, "Why Florida's Manatee Population Is Rebounding," *Christian Science Monitor*, January 16, 2016, https://www.csmonitor.com/Environment/2016/0116/Why-Florida-s-manatee-population-is-rebounding

3. "How Many Species Are We Losing?" World Wildlife Fund, n.d., http://wwf.panda.org/about_our_earth/biodiversity/biodiversity/

4. "Deforestation," Wikipedia, February 28, 2018, https://en.wikipedia.org/wiki/Deforestation

5. Sean Gallagher, "Masked City: The People Who Breathe Beijing's Deadly Air," Mashable, August 15, 2015, https://mashable.com/2015/08/15/masked-city-beijing-air-pollution/#HGrBpa3lIaqr

6. David Jolly, "Norway Is a Model for Encouraging Electric Car Sales," *The New York Times*, October 16, 2015, https://www.nytimes.com/2015/10/17/business/international/norway-is-global-model-for-encouraging-sales-of-electric-cars.html

7. "2018 Outlook for Energy: A View to 2040," ExxonMobil, 2018, http://cdn.exxonmobil.com/~/media/global/files/outlook-for-energy/2018/2018-outlook-for-energy.pdf

8. Oscar van Vliet, Anne Sjoerd Brouwer, Takeshi Kuramochi, and Andre Faaij, "Energy Use, Cost and CO_2 Emissions of Electric Cars," *Journal of Power Sources*, January 2010, https://www.researchgate.net/publication/229376531_Energy_use_cost_and_CO2_emissions_of_electric_cars

9. Reto Bättig and Marco Zielger, "Swiss Greenhouse Gas Abatement Cost Curve," McKinsey & Company, January 2009, https://www.mckinsey.com/~/media/mckinsey/dotcom/client_service/sustainability/cost%20curve%20pdfs/ghg_cost_curve_report_final.ashx

10. "Options for Achieving Deep Reductions in Carbon Emissions in Philadelphia by 2050," A.J. Drexel Institute for Energy and the Environment, November 2015, https://www.researchgate.net/publication/283503711_Options_for_Achieving_Deep_Reductions_in_Carbon_Emissions_in_Philadelphia_by_2050

11. James Sweeney, "A Cost-Effectiveness Analysis of AB 32 Measures," Stanford University Precourt Institute for Energy Efficiency, n.d. https://web.stanford.edu/group/peec/cgi-bin/docs/policy/research/A%20Cost-effectiveness%20Analysis%20of%20AB%2032%20Measures%20(Revised).pdf

12. Gayathri Vaidyanathan, "How Bad of a Greenhouse Gas Is Methane?" *Scientific American*, December 22, 2015, https://www.scientificamerican.com/article/how-bad-of-a-greenhouse-gas-is-methane/

13. "A Fresh Look at the Costs of Reducing US Carbon Emissions," Bloomberg New Energy Finance, January 14, 2010, https://about.

bnef.com/blog/us-mac-curve-a-fresh-look-at-the-costs-of-reducing-us-
carbon-emissions/

14. James Sweeney, "A Cost-Effectiveness Analysis of AB 32 Measures."

15. "Measuring the Cost of a 2°C Target Under the
    Paris Agreement," RepuTex Carbon, May 20th,
    2017, http://www.reputex.com/knowledge-centre/
    article-measuring-the-cost-of-a-2c-target-under-the-paris-agreement/

16. Daniel M. Kammen, Samuel M. Arons, Derek M. Lemoine, and
    Holmes Hummel, "Cost-Effectiveness of Greenhouse Gas Emission
    Reductions from Plug-in Hybrid Electric Vehicles," University of
    California Berkeley Goldman School of Public Policy, n.d., https://gspp.
    berkeley.edu/assets/uploads/research/pdf/ssrn-id1307101.pdf

17. Ross McKitrick, "The High Price of Low Emissions: Benefits and
    Costs of GHG Abatement in the Transportation Sector" Macdonald-
    Laurier Institute, March 2012, https://www.macdonaldlaurier.ca/files/
    pdf/The-high-price-of-low-emissions-benefits-and-costs-of-GHG-
    abatement-in-the-transportation-sector-February-2012.pdf

18. Ross McKitrick, "The High Price of Low Emissions: Benefits and Costs
    of GHG Abatement in the Transportation Sector."

19. Spencer Dale and Thomas D. Smith, "Back to the Future: Electric
    Vehicles and Oil Demand," presentation at Bloomberg New Energy
    Finance: The Future of Energy, EMEA Summit, October 2016, https://
    www.bp.com/en/global/corporate/media/speeches/back-to-the-future-
    electric-vehicles-and-oil-demand.html

20. Akshat Rathi, "A Carbon Tax Killed Coal in the UK.
    Natural Gas Is Next," Quartz, February 1, 2018, https://
    qz.com/1192753/a-carbon-tax-killed-coal-in-the-uk-natural-gas-is-next/

21. Susan A. Shaheen and Timothy E. Lipman, "Reducing Greenhouse
    Emissions and Fuel Consumption: Sustainable Approaches for Surface
    Transportation," IATSS Research, Volume 31, Issue 1, 2007, https://doi.
    org/10.1016/S0386-1112(14)60179-5

22. David L. Green and Steven E. Plotkin, "Reducing Greenhouse Gas
    Emissions from U.S. Transportation," Pew Center on Global Climate
    Change, January 2011, http://cta.ornl.gov/cta/Publications/Reports/
    Reducing_GHG_from_transportation.pdf

23. David L. Green and Steven E. Plotkin, "Reducing Greenhouse Gas
    Emissions from U.S. Transportation."

24. Spencer Dale and Thomas D. Smith, "Back to the Future: Electric
    Vehicles and Oil Demand."

25. David Biello, "Electric Cars Are Not Necessarily Clean."

26. Spencer Dale and Thomas D. Smith, "Back to the Future: Electric Vehicles and Oil Demand."

CHAPTER 5: THE KING IS DEAD. LONG LIVE THE KING.

1. Matt Novak, "Think Inside the Box: What Time Capsules Reveal About Right Now," PaleoFuture, December 9, 2011, https://paleofuture.gizmodo.com/think-inside-the-box-what-time-capsules-reveal-about-r-1669530230

2. Robert Rapier, "What Hubbert Got Really Wrong About Oil," *Forbes*, September 8, 2016, https://www.forbes.com/sites/rrapier/2016/09/08/what-hubbert-got-really-wrong-about-oil/#4fc916b72a3b

3. Michael Lynch, "The Death of the Oil Industry: Not This Again," *Forbes*, February 8, 2017, https://www.forbes.com/sites/michaellynch/2017/02/08/the-death-of-the-oil-industry-not-this-again/

4. "FAQs: Oil," International Energy Agency, n.d., https://www.iea.org/about/faqs/oil/

5. Michael Lynch, "The Death of the Oil Industry: Not This Again."

6. Ray Prince, "Barclays Report: Oil Demand Will Plummet by 2025 Due to Electric Cars," October 6, 2017, hybridcars.com, http://www.hybridcars.com/barclays-report-oil-demand-will-plummet-by-2025-due-to-electric-cars/

7. "FAQs: Oil," International Energy Agency.

8. Harry Goldstein and William Sweet, "Joules, BTUs, Quads—Let's Call the Whole Thing Off," *IEEE Spectrum*, January 1, 2007, https://spectrum.ieee.org/energy/fossil-fuels/joules-btus-quads-lets-call-the-whole-thing-off

9. "Cubic Mile of Oil," Wikipedia, November 3, 2017, https://en.wikipedia.org/wiki/Cubic_mile_of_oil

10. Robert Lyman, "Why Renewable Energy Cannot Replace Fossil Fuels by 2050," Friends of Science, May 2016, https://www.friendsofscience.org/assets/documents/Renewable-energy-cannot-replace-FF_Lyman.pdf

11. Robert Lyman, "Why Renewable Energy Cannot Replace Fossil Fuels by 2050."

12. Robert Bryce, "Don't Count Oil Out," *Slate*, October 14, 2011, http://www.slate.com/articles/technology/future_tense/2011/10/oil_and_gas_won_t_be_replaced_by_alternative_energies_anytime_so.html

13. Jane Burgermeister, "Germany: The World's First Major Renewable Energy Economy," Renewable Energy World, April 3, 2009, http://www.renewableenergyworld.com/articles/2009/04/germany-the-worlds-first-major-renewable-energy-economy.html

14. Jesper Starn, "Old Coal Is King Even with New Renewables Record in Germany," Bloomberg, December 17, 2017, https://www.bloomberg.com/news/articles/2017-12-18/old-coal-is-king-even-with-another-renewables-record-in-germany

15. Kerstine Appunn, Felix Bieler, and Julian Wettengel, "Germany's Energy Consumption and Power Mix in Charts," Clean Energy Wire, February 16, 2018, https://www.cleanenergywire.org/factsheets/germanys-energy-consumption-and-power-mix-charts

16. Sören Amelang, "Germany on Track to Widely Miss 2020 Climate Target," Clean Energy News, June 13, 2018, https://www.cleanenergywire.org/news/germany-track-widely-miss-2020-climate-target-government

17. Kerstine Appunn, "Germany's Greenhouse Gas Emissions and Climate Targets," Clean Energy Wire, February 1, 2018, https://www.cleanenergywire.org/factsheets/germanys-greenhouse-gas-emissions-and-climate-targets

18. James Conca, "Why Aren't Renewables Decreasing Germany's Carbon Emissions?" Forbes, October 10, 2017, https://www.forbes.com/sites/jamesconca/2017/10/10/why-arent-renewables-decreasing-germanys-carbon-emissions/#6d0fb8be68e1

19. Dave Jones, "European Coal Emissions Decline but German Lignite Stations Keep Pumping On," The Energy Collective, April 4, 2017, http://www.theenergycollective.com/dave-jones/2401723/european-coal-emissions-decline-german-lignite-stations-keep-pumping

20. "2017 Outlook for Energy: A View to 2040," ExxonMobil, 2017, http://cdn.exxonmobil.com/~/media/global/files/outlook-for-energy/2017/2017-outlook-for-energy.pdf

21. Robert Bryce, Power Hungry: The Myths of "Green" Energy and the Real Fuels of the Future, New York: Business News Publishing, January 30, 2017, https://books.google.com/books?id=aAOBjN97xqAC&printsec=frontcover&source=gbs_ge_summary_r&cad=0#v=onepage&q&f=false.

22. "Carbon Tracker Initiative," Energy Transition Advisors, May 8, 2014, http://www.carbontracker.org/wp-content/uploads/2014/05/Chapter1ETAdemandfinal.pdf

23. Simon Göß, "Power Statistics China 2016: Huge Growth of Renewables

Amidst Thermal-Based Generation," *Energy BrainBlog*, February 9, 2017, https://blog.energybrainpool.com/en/power-statistics-china-2016-huge-growth-of-renewables-amidst-thermal-based-generation

24. "Worldwide number of battery electric vehicles in use from 2012 to 2017," Statista, 2019, https://www.statista.com/statistics/270603/worldwide-number-of-hybrid-and-electric-vehicles-since-2009/

25. "Worldwide Car Sales 1990-2019," Statista, 2019, https://www.statista.com/statistics/200002/international-car-sales-since-1990/

26. "2017 Outlook for Energy: A View to 2040," ExxonMobil.

27. Cuneyt Kazokoglu, "Electric Cars Pose Little Threat to Oil Demand," *Financial Times*, March 21, 2017, https://www.ft.com/content/502c4e3c-0d80-11e7-b030-768954394623

28. "China FAQs: Coal for Electricity," The Network for Climate and Energy Information Convened by the World Resources Institute, n.d., http://www.chinafaqs.org/issue/coal-electricity

29. "BP Energy Outlook: 2019 Edition," BP, February 14, 2019, https://www.bp.com/content/dam/bp/business-sites/en/global/corporate/pdfs/energy-economics/energy-outlook/bp-energy-outlook-2019.pdf

30. "International Energy Outlook 2017," U.S. Energy Information Administration, September 14, 2017, https://www.eia.gov/outlooks/ieo/pdf/0484(2017).pdf

31. "Carbon Tracker Initiative," Energy Transition Advisors.

32. "New Lens Scenarios: A Shift in Perspective for a World in Transition," Shell, March 2013, https://www.shell.com/energy-and-innovation/the-energy-future/scenarios/new-lenses-on-the-future/_jcr_content/par/relatedtopics.stream/1448477051486/08032d761ef7d81a4d3b1b6df8620c1e9a64e564a9548e1f2db02e575b00b765/scenarios-newdoc-english.pdf

33. "International Energy Outlook 2017," U.S. Energy Information Administration.

34. Tom Randall, "The Electric-Car Boom Is So Real Even Oil Companies Say It's Coming," Bloomberg, April 25, 2017, https://www.bloomberg.com/news/articles/2017-04-25/electric-car-boom-seen-triggering-peak-oil-demand-in-2030s

35. Jess Shankleman, "Big Oil Just Woke Up to Threat of Rising Electric Car Demand," Bloomberg, July 14, 2017, https://www.bloomberg.com/news/articles/2017-07-14/big-oil-just-woke-up-to-the-threat-of-rising-electric-car-demand

36. Nawar Alsaadi, "The EV Myth—Electric Car Threat to Oil Is Wildly Overstated," OilPrice.com, March 2, 2017, https://oilprice.com/Energy/Energy-General/The-EV-Myth-Electric-Car-Threat-To-Oil-Is-Wildly-Overstated.html

37. "2017 Outlook for Energy: A View to 2040."

38. Scott Shepard, "Here's How Electric Cars Will Not Cause the Next Oil Price Crash," Navigant Research, March 21, 2016, https://www.navigantresearch.com/blog/heres-how-electric-cars-will-not-cause-the-next-oil-price-crash

CHAPTER 6: ARE SUBSIDIES "LUDICROUS"?

1. Anton Wahlman, "Elon Musk Begs The Feds: Please End Tesla's Tax Subsidy," Seeking Alpha, May 4, 2017, https://seekingalpha.com/article/4069065-elon-musk-begs-feds-please-end-teslas-tax-subsidy

2. "Advanced Technology Vehicles Manufacturing Loan Program," Wikipedia, November 6, 2017, https://en.wikipedia.org/wiki/Advanced_Technology_Vehicles_Manufacturing_Loan_Program

3. Allen R. Myerson, "O Governor, Won't You Buy Me a Mercedes Plant?," *The New York Times*, September 1, 1996, http://www.nytimes.com/1996/09/01/business/o-governor-won-t-you-buy-me-a-mercedes-plant.html

4. "Vernon, Alabama," Wikipedia, January 12, 2018, https://en.wikipedia.org/wiki/Vernon,_Alabama

5. "Energy Independence and Security Act (EISA) of 2007" (full text), Government Publishing Office, December 19, 2007, https://www.gpo.gov/fdsys/pkg/STATUTE-121/pdf/STATUTE-121-Pg1492.pdf

6. "Obama Administration Awards First Three Auto Loans for Advanced Technologies to Ford Motor Company, Nissan Motors and Tesla Motors," U.S. Department of Energy, June 23, 2009, https://energy.gov/articles/obama-administration-awards-first-three-auto-loans-advanced-technologies-ford-motor-company

7. Chris Isidore, "Chrysler Jabs Tesla Over Loan Repayment," CNN Money, May 23, 2013, http://money.cnn.com/2013/05/23/news/companies/tesla-chrysler-spat/index.html

8. Anton Wahlman, "Elon Musk Begs The Feds: Please End Tesla's Tax Subsidy."

9. Trefis Team, "Tesla's Lucrative ZEV Credits May Not Be Sustainable," *Forbes*, September 1, 2017, https://

www.forbes.com/sites/greatspeculations/2017/09/01/
teslas-lucrative-zev-credits-may-not-be-sustainable/#120d0a6b6ed5

10. Danielle Muoio, "Elon Musk Says Trump Presidency Won't Hurt
    Tesla—Here's Why," *Business Insider*, November 20, 2016, http://
    www.businessinsider.com/elon-musk-trump-decision-on-electric-car-
    incentives-wont-hurt-tesla-2016-11

11. Anton Wahlman, "Elon Musk Begs The Feds: Please End Tesla's Tax
    Subsidy."

12. "US Vice President Joe Biden Inadvertently
    Divulges Fisker Product Plan," Motor1, October
    30, 2009, https://www.motor1.com/news/18452/
    us-vice-president-joe-biden-inadvertantly-divulges-fisker-product-plan/

13. Matthew Mosk, Brian Ross, and Ronnie Greene, "Car
    Company Gets U.S. Loan, Builds Cars In Finland," ABC
    News, October 20, 2011, https://abcnews.go.com/Blotter/
    car-company-us-loan-builds-cars-finland/story?id=14770875

14. Richard Lawler, "Fisker Karma Recall Is Official, 239
    Cars Will Need Their Battery Packs Swapped," *Engadget*,
    January 1, 2012, https://www.engadget.com/2012/01/01/
    fisker-karma-recall-is-official-239-cars-will-need-their-batter/

15. "Bad Karma: Our Fisker Karma Plug-in Hybrid Breaks Down,"
    Consumer Reports News, March 8, 2012, https://www.
    consumerreports.org/cro/news/2012/03/bad-karma-our-fisker-karma-
    plug-in-hybrid-breaks-down/index.htm

16. "A123Systems Awarded $249M Grant from U.S. Department
    of Energy to Build Advanced Battery Production Facilities in
    the United States," Business Wire, August 5, 2009, https://
    www.businesswire.com/news/home/20090805005904/en/
    A123Systems-Awarded-249M-Grant-U.S.-Department-Energy

17. Peter W. Davidson, "An Update on Fisker Automotive and
    the Energy Department's Loan Portfolio," U.S. Department
    of Energy, September 17, 2013, https://energy.gov/articles/
    update-fisker-automotive-and-energy-department-s-loan-portfolio

18. Deepa Seetharaman and Paul Lienert, "Special Report: Bad
    Karma: How Fisker Burned Through $1.4 Billion on a 'Green'
    Car," Reuters, June 17, 2013, https://www.reuters.com/article/
    us-autos-fisker-specialreport/special-report-bad-karma-how-fisker-
    burned-through-1-4-billion-on-a-green-car-idUSBRE95G02L20130617

19. Nicolas Loris, "Green Energy Oversight: Examining the Department
    of Energy's Bad Bet on Fisker Automotive," Testimony before the

Committee on Oversight and Government Reform Subcommittee
on Economic Growth, Job Creation, and Regulatory Affairs, April 24,
2013, https://oversight.house.gov/wp-content/uploads/2013/04/LORIS-
Testimony.pdf

20. D'Angelo Gore, "Obama's Solyndra Problem," FactCheck.org, October 7,
2011, https://www.factcheck.org/2011/10/obamas-solyndra-problem/

21. Joe Stephens and Carol Leonnig, "Solyndra: Politics Infused
Obama Energy Programs," *The Washington Post*, December
25, 2011, https://www.washingtonpost.com/solyndra-politics-
infused-obama-energy-programs/2011/12/14/gIQA4HllHP_story.
html?utm_term=.7019d8da014e

22. "Special Report: The Department of Energy's Loan Guarantee to
Solyndra, Inc.," U.S. Department of Energy Office of Inspector General,
August 24, 2015, https://energy.gov/sites/prod/files/2015/08/f26/11-
0078-I.pdf

23. "World Trade Report 2006," World Trade Organization, 2006,
https://www.wto.org/english/res_e/booksp_e/anrep_e/world_trade_
report06_e.pdf

24. Matthew Campbell, "The Country Adopting Electric Vehicles
Faster Than Anywhere Else," Bloomberg Businessweek, May 31,
2017, https://www.bloomberg.com/news/articles/2017-06-01/
the-country-adopting-electric-vehicles-faster-than-anywhere-else

25. Jamie Merrill, "Saving Money Not the Environment Is Driving Norway's
Electric Car Boom," *The Independent*, June 14, 2014, http://www.
independent.co.uk/life-style/motoring/motoring-news/saving-money-
not-the-environment-is-driving-norways-electric-car-boom-9537737.html

26. David Yager, "Electric Vehicles No Threat to Oil Prices Anytime Soon,"
OilPrice.com, July 27, 2017, https://oilprice.com/Alternative-Energy/
Renewable-Energy/Electric-Vehicles-No-Threat-To-Oil-Prices-Anytime-
Soon.html

27. "Production Forecasts," Norwegian Petroleum, March 26, 2019,
https://www.norskpetroleum.no/en/production-and-exports/
production-forecasts/

28. "The Government's Revenues," Norwegian Petroleum, March 19, 2019,
http://www.norskpetroleum.no/en/economy/governments-revenues/

29. Matthew Campbell, "The Country Adopting Electric Vehicles Faster
Than Anywhere Else."

30. Richard Milne, "Reality of Subsidies Drives Norway's Electric

Car Dream, *Financial Times*, June 14, 20174, https://www.ft.com/content/84e54440-3bc4-11e7-821a-6027b8a20f23

31. Michael Lynch, "Explaining the Appeal of Energy Revolutions (That Ultimately Failed)," *Forbes*, June 12, 2017, https://www.forbes.com/sites/michaellynch/2017/06/12/explaining-the-appeal-of-energy-revolutions-that-ultimately-failed/2/#346887185ec2

32. "ATVM Loan Program," US Department of Energy Loan Programs Office, n.d., https://www.energy.gov/lpo/services/atvm-loan-program

CHAPTER 7: CHINA IN THE DRIVER'S SEAT—OR NOT

1. Jack Perkowski, "What China's Shifting Subsidies Could Mean For Its Electric Vehicle Industry," *Forbes*, July 13, 2018, https://www.forbes.com/sites/jackperkowski/2018/07/13/china-shifts-subsidies-for-electric-vehicles/#611ce6695703

2. Robert Rapier, "China Emits More Carbon Dioxide Than the U.S. and EU Combined," *Forbes*, July 1, 2018, https://www.forbes.com/sites/rrapier/2018/07/01/china-emits-more-carbon-dioxide-than-the-u-s-and-eu-combined/#4b0bab10628c

3. Matthew Campbell and Tian Ying, "China Is Leading the World to an Electric Car Future," Bloomberg Businessweek, November 14, 2018, https://www.bloomberg.com/news/articles/2018-11-14/china-is-leading-the-world-to-an-electric-car-future

4. Gabriel Collins, "China's Evolving Oil Demand," Rice University's Baker Institute for Public Policy, 2016, https://www.bakerinstitute.org/media/files/files/e0b5a496/WorkingPaper-ChinaOil-093016.pdf

5. Trefor Moss, "Global Auto Makers Dented as China Car Sales Fall for First Time in Decades," *The Wall Street Journal*, January 14, 2019, https://www.wsj.com/articles/chinese-annual-car-sales-slip-for-first-time-in-decades-11547465112

6. "Focus on China—2015 Market Overview: Impacts and Opportunities of Car Purchase Restrictions," JATO Dynamics, March 2016, http://www.jato.com/wp-content/uploads/2016/04/JATO-Market-Focus-China-Car-Purchase-restrictions-and-Market-overview-2015-Final.pdf

7. "China to Cut New Coal-Fired Power Plants in 29 Provinces," *The Sydney Morning Herald*, May 14, 2017, https://www.smh.com.au/business/china-to-cut-new-coalfired-power-plants-in-29-provinces-20170512-gw3lj5.html

8. Gabriel Collins, "China's Evolving Oil Demand," Baker Institute,

September 30, 2016, https://www.bakerinstitute.org/research/chinas-evolving-oil-demand/

9. Danial Dzulkifly, "Electric Cars in Malaysia not so Green After All, Study Shows as Govt Plans Third National Car," *Malay Mail*, August 24, 2018, https://www.malaymail.com/news/malaysia/2018/08/24/electric-cars-in-malaysia-not-so-green-after-all-study-shows-as-govt-plans/1665532

10. Stephen Fogel, "EV Market Profile: Vietnam," What Auto, https://whatauto.expert/ev-market-profile-vietnam/

11. "Made in China?" *The Economist*, March 12, 2015, https://www.economist.com/leaders/2015/03/12/made-in-china

12. Alaric Nightingale, "Forget Tesla, It's China's E-Buses That Are Denting Oil Demand," Bloomberg, March 19, 2019, https://www.bloomberg.com/news/articles/2019-03-19/forget-tesla-it-s-china-s-e-buses-that-are-denting-oil-demand

13. Natalie Sauer, "Electric Cars 'Won't Stop Rising Oil Demand'," *The Ecologist*, January 25, 2019, https://theecologist.org/2019/jan/25/electric-cars-wont-stop-rising-oil-demand

14. Marianne Kah, "Electric Vehicles and Their Impact on Oil Demand: Why Forecasts Differ," Columbia University Center in Global Energy Policy July 24, 2018, https://energypolicy.columbia.edu/research/commentary/electric-vehicles-and-their-impact-oil-demand-why-forecasts-differ

15. Isabelle Niu, "Your Next Car Could Be Electric—and Chinese," Quartz, November 15, 2018, https://qz.com/1463563/your-next-car-could-be-electric-and-chinese/

16. Jack Perkowski, "What China's Shifting Subsidies Could Mean for its Electric Vehicle Industry," *Forbes*, July 13, 2018, https://www.forbes.com/sites/jackperkowski/2018/07/13/china-shifts-subsidies-for-electric-vehicles/#6f93c1f85703

17. Ying Tian, Yan Zhang, and Jie Ma, "The $18 Billion Electric-Car Bubble at Risk of Bursting in China," Bloomberg, April 16, 2019, https://governorswindenergycoalition.org/the-18-billion-electric-car-bubble-at-risk-of-bursting-in-china/

18. David Reid, "China's Electric Vehicle Sales Will Continue Boom Despite Subsidy Cuts, Fitch Says," CNBC, April 8, 2019, https://www.cnbc.com/2019/04/08/fitch-says-chinese-electric-vehicle-sales-to-boom-despite-subsidy-cuts.html

19. Gabriel Collins, "China's Evolving Oil Demand."

20. Angaindrankumar Gnanasagaran, "Southeast Asia's Electric Car Revolution," *The Asean Post*, November 27, 2018, https://theaseanpost. com/article/southeast-asias-electric-car-revolution

21. Pamela Victor, "Thailand's Electric Vehicle Dreams," *The Asean Post*, February 18, 2018, https://theaseanpost.com/article/ thailands-electric-vehicle-dreams

22. Echo Huang, "Nobody in Hong Kong Wants a Tesla Anymore," Quartz, July 10, 2017,https://qz.com/1024886/ nobody-in-hong-kong-wants-a-teslanasdaq-tslaanymore/

23. Fred Lambert, "Hong Kong Brings Back Some Electric Vehicle Incentives that Made Tesla so Popular in the Region," Electrek, February 28, 2018, https://electrek.co/2018/02/28/ hong-kong-electric-cars-incentives-tesla/

24. Wade Shepard, "Why Chinese Cities Are Banning The Biggest Adoption Of Green Transportation In History," *Forbes*, May 18, 2016, https://www.forbes.com/sites/wadeshepard/2016/05/18/ as-china-chokes-on-smog-the-biggest-adoption-of-green-transportation- in-history-is-being-banned/#327e1482141b

25. Qian Zecheng, "China to Roll Out Stricter Standards for Electric Bikes," *Sixth Tone*, January 17, 2018, https://www.sixthtone.com/ news/1001569/china-to-roll-out-stricter-standards-for-electric-bikes

26. Trefor Moss, "China's Giant Market for Really Tiny Cars," *The Wall Street Journal*, September 21, 2018, https://www.wsj.com/articles/ chinas-giant-market-for-tiny-cars-1537538585

PART II: THE CARS
EV OPTIONS AND OWNERSHIP

CHAPTER 8: EV AUTOMAKERS: MAKING IT OR BREAKING IT

1. Noel McKeegan, "Aptera Officially Launches Futuristic, Super-Efficient Three Wheeler," *New Atlas*, November 23, 2007, https://newatlas.com/ aptera-typ1-three-wheel-electric-vehicle/8392

2. John Voelcker, "Aptera Collapse: How & Why It Happened, A Complete Chronology," Green Car Reports, December 12, 2011, https://www. greencarreports.com/news/1070490_aptera-collapse-how-why-it- happened-a-complete-chronology

3. Chris Woodyard, "Drive On: Coda Becomes Latest Electric Car

Failure," *USA Today*, May 1, 2013, https://www.usatoday.com/story/
driveon/2013/05/01/coda-bankruptcy-electric-car/2127673

4. "Electrifying Insights: How Automakers Can Drive Electrified Vehicle
Sales and Profitability," McKinsey & Co., January 2017, https://www.
mckinsey.com/industries/automotive-and-assembly/our-insights/
electrifying-insights-how-automakers-can-drive-electrified-vehicle-sales-
and-profitability

5. Michael Lynch, "What Keeps Killing Electric Cars?" *Forbes*, April 8,
2016,

6. Trent Eady, "The Model 3 Can Make Tesla Profitable,"
Seeking Alpha, January 22, 2018, https://seekingalpha.com/
article/4139162-model-3-can-make-tesla-profitable

7. Cadie Thompson, "Tesla's Model 3 Has Already Dramatically
Changed the Automotive Industry—Here's How," *Business
Insider*, July 29, 2017, http://www.businessinsider.com/
elon-musk-electric-cars-plan-is-working-2017-7

8. Aaron Brown, "Here's a Look at the Tesla Car that Started It All,"
Business Insider, March 30, 2016, http://www.businessinsider.com/
tesla-roadster-history-2016-3

9. Davey G. Johnson, "2012 Tesla Model S: Drive Review: The Most Fully
Realized EV on the Market Is a (Very Quiet) Hoot," *Autoweek*, June 24,
2012, http://autoweek.com/article/car-reviews/2012-tesla-model-s-drive-
review-most-fully-realized-ev-market-very-quiet-hoot#ixzz57kis5qJp

10. Tony Quiroga, "2016 Tesla Model X," *Car and Driver*, May 2016, https://
www.caranddriver.com/reviews/2016-tesla-model-x-test-review

11. Nick Lucchesi, "The Tesla Model 3 Delay Timeline: How It Got
to 6 Months," Inverse, January 3, 2018, https://www.inverse.com/
article/39907-tesla-model-3-delay-january-2018

12. "Tesla Net Worth 2009-2020," Macrotrends, May 2020, https://www.
macrotrends.net/stocks/charts/TSLA/tesla/net-worth

13. Ross Gerber, "Tesla's Drive to Save the World and
Grow to the Trees," *Forbes*, February 8, 2017, https://
www.forbes.com/sites/greatspeculations/2017/02/08/
teslas-drive-to-save-the-world-and-grow-to-the-trees/#2fc23920647d

14. Michael Sheetz, "Tesla's Charging Stations Are a Massive 'Competitive
Moat,' Morgan Stanley Says," CNBC, February 12, 2019, https://www.
cnbc.com/2019/02/12/morgan-stanley-tesla-charging-station-network-
competitive-moat.html

15. Fred Lambert, "Tesla's Latest Software Update Brings More Supercharger Information," Electrek, September 14, 2017, https://electrek.co/2017/09/14/tesla-software-update-supercharging-information/

16. Sean Szymkowski, "Fiat-Chrysler Chief Sergio Marchionne Doesn't Think Electric Cars, Tesla Are Viable," Green Car Reports, October 12, 2017, https://www.greencarreports.com/news/1113204_fiat-chrysler-chief-sergio-marchionne-doesnt-think-electric-cars-tesla-are-viable

17. Aarian Marshall, "Ford Finally Makes Its Move into Electric Cars," *Wired*, January 17, 2018, https://www.wired.com/story/ford-electric-cars-plan-mach-1-suv

18. "Monthly Plug-In Sales Scorecard," InsideEVs, January 2018, https://insideevs.com/monthly-plug-in-sales-scorecard

19. JC Reindl, "This $750K Converted, All-Electric Corvette Could Hit 220 m.p.h." *Detroit Free Press*, January 14, 2018, https://www.freep.com/story/money/cars/general-motors/2018/01/14/converted-electric-corvette/1026956001/

20. Eric C. Evarts, "Lucid Motors Gets Real as Saudi Funding Comes Through," Green Car Reports, April 5, 2019, https://www.greencarreports.com/news/1122459_lucid-motors-gets-real-as-saudi-funding-comes-through

21. Bryan Campbell, "Supercar vs Hypercar—What's the Difference?" Gear Patrol, n.d., https://gearpatrol.com/2018/01/16/supercar-hypercar-defined-what-is-a-supercar/

22. Nick Hall, "Everything You Need to Know About the Rimac Concept One," Dgit, August 8, 2017, https://dgit.com/rimac-concept-one-power-speed-range-torque-price-983/

23. Sean O'Kane, "Byton's Electric SUV Concept Is Another Wild Stab at the Future of Cars," The Verge, January 7, 2018, https://www.theverge.com/2018/1/7/16859490/byton-electric-car-concept-suv-price-range-speed-ces-2018

24. Fred Lambert, "Tesla-Inspired Chinese EV Startup Launches All-Electric SUV Using Open-Source Patents," Electrek, October 24, 2017, https://electrek.co/2017/10/24/tesla-clone-chinese-ev-startup-all-electric-suv-open-source-patent/

25. Erin Griffith, "In Building an Electric Car, Dyson Goes Its Own Way," *Wired*, October 4, 2017, https://www.wired.com/story/in-building-an-electric-car-dyson-goes-its-own-way/

CHAPTER 9: WHAT CONSUMERS DON'T LIKE (OR THINK THEY DON'T)

1. John Voelcker, "The Challenge for Electric-Car Sales Is Car Dealers, Again," Green Car Reports, August 5, 2014, https://www.greencarreports.com/news/1093687_the-challenge-for-electric-car-sales-is-car-dealers-again

2. Justin Gerdes, "US Auto Dealerships Are Bad at Selling Electric Vehicles, Study Finds," Greentech Media, December 12, 2017, https://www.greentechmedia.com/articles/read/us-auto-dealerships-are-bad-at-selling-electric-vehicles-study-finds#gs.UiLMKhk

3. Justin Gerdes, "US Auto Dealerships Are Bad at Selling Electric Vehicles, Study Finds."

4. John Voelcker, "2017 Chevy Bolt EV Electric Car: New Owner's First Impressions," Green Car Reports, February 6, 2017, https://www.greencarreports.com/news/1108723_2017-chevy-bolt-ev-electric-car-new-owners-first-impressions/page-3

5. "Consumer Reviews: 2017 Chevrolet Bolt EV," Edmunds, n.d., https://www.edmunds.com/chevrolet/bolt-ev/2017/consumer-reviews/pg-1/?sorting=CONFIDENCE

6. Bill Howard, "How Do Electric Cars Work," ExtremeTech, July 6, 2015, https://www.extremetech.com/extreme/209190-tech-backgrounder-the-merits-and-challenges-of-electric-cars

7. Antony Ingram, "Do EVs Need a Passing Gear? Nissan Says Yes, Chevy Disagrees," Green Car Report, July 13, 2010, https://www.greencarreports.com/news/1047135_do-evs-need-a-passing-gear-nissan-says-yes-chevy-disagrees

8. Elizabeth Stinson, "EVs Are Dangerously Quiet. Here's What They Could Sound Like," Wired, April 3, 2017, https://www.wired.com/2017/04/evs-dangerously-quiet-heres-sound-like/

9. Dave Chameides, "12 Myths About Electric Vehicles," How Stuff Works, December 6, 2011, https://auto.howstuffworks.com/myths-electic-cars-vehicles1.htm

10. Fred Lambert, "Tesla Releases New Longer Range Model S and Model X with Drivetrain, Suspension Upgrades & More," Electrek, April 23, 2019, https://electrek.co/2019/04/23/tesla-new-model-s-x-range-upgrade-drivetrain-suspension/

11. "Consumer Reviews: 2017 Chevrolet Bolt EV."

12. Eric Taub, "For Electric Car Owners, 'Range Anxiety' Gives Way to

'Charging Time Trauma,'" *The New York Times*, October 5, 2017, https://www.nytimes.com/2017/10/05/automobiles/wheels/electric-cars-charging.html

13. Norman Mayersohn, "The Internal Combustion Engine Is Not Dead Yet," *The New York Times*, August 12, 2017, https://www.nytimes.com/2017/08/17/automobiles/wheels/internal-combustion-engine.html

14. John Voelcker, "2017 Chevy Bolt EV Electric Car: New Owner's First Impressions."

15. Nikki Gordon-Bloomfield, "Life in the Freezer: How a Chevy Volt Drives in a Hard Canadian Winter," Transport Evolved, January 2, 2014, https://transportevolved.com/2014/01/02/life-in-the-freezer-how-a-chevy-volt-drives-in-a-hard-canadian-winter/

16. Eric Taub, "For Electric Car Owners, 'Range Anxiety' Gives Way to 'Charging Time Trauma.'"

17. "Alternative Fueling Station Locator," US Department of Energy Office of Energy Efficiency & Renewable Energy, May 2019, https://afdc.energy.gov/stations/#/analyze?country=US&fuel=ELEC&ev_levels=all

18. Paul Ausick, "Why Are There 115,000 (or 150,000) Gas Stations in America?" 24/7 Wall St., May 22, 2014, http://247wallst.com/economy/2014/05/22/why-are-there-115000-or-150000-gas-stations-in-america/

19. Julie Hall, "Family Vacations Still a Popular Priority for Millions of Americans," AAA NewsRoom, February 7, 2017, http://newsroom.aaa.com/2017/02/family-vacations-still-popular-priority-millions-americans/

20. Bengt Halvorson, "Electric-Car Road Trips: Be Prepared, with Charging Apps and Realism," Green Car Reports, December 10, 2014, https://www.greencarreports.com/news/1095840_electric-car-road-trips-be-prepared-with-charging-apps-and-realism

21. "How Long Does It Take to Charge an Electric Car?" Pod Point, n.d., https://pod-point.com/landing-pages/how-long-does-it-take-to-charge-an-electric-car#charge-time

22. Eric Schaal, "5 Ways Electric Vehicle Charging Can Improve," FleetCarma, June 28, 2016, https://www.fleetcarma.com/5-ways-electric-vehicle-charging-can-improve/

23. John Voelcker, "2017 Chevrolet Volt Review," The Car Connection, June 5, 2017, https://www.thecarconnection.com/overview/chevrolet_volt_2017

24. Yuliya Chernova, "Who Drives Electric Cars," *The Wall Street Journal*, September 23, 2013, https://www.wsj.com/articles/ who-drives-electric-cars-1379884645

25. Mark Kane, "1,000,000 Plug-in Electric Cars Sold in US," InsideEVs, October 6, 2018, https://insideevs.com/news/340135/ plug-in-electric-cars-sales-in-us-surpass-1-million/

26. "Pricing," ecar, n.d., https://ecarclub.co.uk/pricing/

27. Michael J. Coren, "Electric Vehicle Car Clubs Let Hesitant Drivers Get a Taste of a Gas-Free Life," *Fast Company*, October 24, 2012, https://www.fastcompany.com/2680773/ electric-vehicle-car-clubs-let-hesitant-drivers-get-a-taste-of-a-gas-free-life

28. "Ride and Drives," Drive Electric Northern Colorado, n.d., http:// driveelectricnoco.org/ride-and-drives/

29. Lingzhi Jin and Peter Slowik, "Literature Review of Electric Vehicle Consumer Awareness and Outreach Activities," International Council on Clean Transportation, March 21, 2017, https://www.theicct.org/sites/default/files/publications/ Consumer-EV-Awareness_ICCT_Working-Paper_23032017_vF.pdf

30. "Electric Vehicle Tourism in Oregon Wine Country!" Oregon Wine Country Electric Vehicle (EV) Byway and Plug & Pinot, n.d., http://www. plugandpinot.com/

31. Matthew Klippenstein, "E-Mazing Race III: Electric Cars Compete on Recharging In Multiple Locations," Green Car Reports, May 29, 2015, https://www.greencarreports.com/news/1098497_e-mazing-race-iii- electric-cars-compete-on-recharging-in-multiple-locations

CHAPTER 10: COST AND SAVINGS: HYPE VERSUS REALITY

1. Douglas Elbinger, "One Year with My Chevy Volt," EV Obsession, July 14, 2013, https://evobsession.com/chevy-volt-owner-review-after-1-year/

2. "Consumer Reviews: 2017 Chevrolet Bolt EV," Edmunds, n.d., https:// www.edmunds.com/chevrolet/bolt-ev/2017/consumer-reviews/ pg-1/?sorting=CONFIDENCE

3. U.S. Department of Transportation Federal Highway Administration, "Average Annual Miles per Driver by Age Group," July 13, 2016, https:// www.fhwa.dot.gov/ohim/onh00/bar8.htm

4. "Complete Transmission Repair Cost Guide," n.d., https://www. transmissionrepaircostguide.com/

5. Christopher Lampton, "How Regenerative Braking Works," How Stuff

Works, n.d., https://auto.howstuffworks.com/auto-parts/brakes/brake-types/regenerative-braking1.htm

6. Steve Hanley, "10 Myths About Electric Cars. Are Any of Them True?" Gas2, October 16, 2017, https://gas2.org/2017/10/16/10-myths-about-electric-cars-any-true/

7. John Voelcker, "Nissan Leaf New Battery Cost: $5,500 for Replacement with Heat-Resistant Chemistry," Green Car Reports, June 28, 2014, http://www.greencarreports.com/news/1092983_nissan-leaf-battery-cost-5500-for-replacement-with-heat-resistant-chemistry

8. Idaho Testing Laboratory, "How Do Gasoline & Electric Vehicles Compare?" n.d.,https://avt.inl.gov/sites/default/files/pdf/fsev/compare.pdf

9. Mark Clayton, "My Nissan Leaf Life: Why I Bought a Nissan Leaf," Christian Science Monitor, May 1, 2012, https://www.csmonitor.com/Business/In-Gear/2012/0501/My-Nissan-Leaf-life-Why-I-bought-a-Nissan-Leaf

10. "Prices of Used Plug-in Electric Vehicles to Drop 30% in 2013," National Automobile Dealers Association, June 19, 2013, https://www.nada.org/CustomTemplates/DetailPressRelease.aspx?id=21474839590

11. Julie Blackley, "Cars With the Lowest and Highest Depreciation," iSeeCars, n.d., https://www.iseecars.com/cars-low-high-depreciation-2018-study

12. Sasha Lekach, "End of an era: 2020 brings Tesla's federal tax credit to zero," December 26, 2019, https://mashable.com/article/tesla-federal-ev-tax-credit-runs-out-2020/

13. "Counting the Cost of Electric Cars," CoverHound, April 22, 2014, https://coverhound.com/insurance-learning-center/counting-the-cost-of-electric-cars

14. NerdWallet, "How Hybrid and Electric Vehicles Affect Your Auto Insurance Quotes," April 1, 2015, https://www.nerdwallet.com/blog/insurance/auto-insurance-quotes-hybrid-electric-cars/

15. Douglas Elbinger, "One Year with My Chevy Volt," EV Obsession, July 14, 2013, https://evobsession.com/chevy-volt-owner-review-after-1-year/

16. Damian Carrington, "Electric Cars 'Will Be Cheaper than Conventional Vehicles by 2022,'" The Guardian, February 25, 1016, https://www.theguardian.com/environment/2016/feb/25/electric-cars-will-be-cheaper-than-conventional-vehicles-by-2022

PART III: THE DETAILS
CHARGING, BATTERIES, ROADS, AND . . . PLASTICS?
CHAPTER 11: INFRASTRUCTURE: IF YOU BUILD IT,
THEY WILL COME

1. Robert Siegel and Andrea Hsu, "In America's Heartland, a Power Company Leads Charge for Electric Cars," NPR, February 14, 2017, https://www.npr.org/sections/alltechconsidered/2017/02/14/514517425/in-americas-heartland-a-power-company-leads-charge-for-electric-cars

2. "Electric Vehicles in Kansas," Plug In America, n.d., https://pluginamerica.org/wp-content/uploads/2017/04/Kansas_Electric_Vehicle_Factsheet_May_2017.pdf

3. Robert Siegel and Andrea Hsu, "In America's Heartland, a Power Company Leads Charge for Electric Cars."

4. Martha T. Moore, "Should Utilities Build Charging Stations for Electric Cars?" The Pew Charitable Trusts, September 11, 2017, http://www.pewtrusts.org/en/research-and-analysis/blogs/stateline/2017/09/11/should-utilities-build-charging-stations-for-electric-cars

5. Robert Siegel and Andrea Hsu, "In America's Heartland, a Power Company Leads Charge for Electric Cars."

6. Martha T. Moore, "Should Utilities Build Charging Stations for Electric Cars?"

7. Martha T. Moore, "Should Utilities Build Charging Stations for Electric Cars?"

8. "Power Your Drive," San Diego Gas & Electric Company, n.d., https://www.sdge.com/clean-energy/electric-vehicles/poweryourdrive

9. "The Valley Hospital and PSE&G Partner on Electric Vehicle Charging System and Energy Efficiency Improvements," Public Service Enterprise Group Newsroom, March 29, 2016, https://www.pseg.com/info/media/newsreleases/2016/2016-03-29.jsp#.Wn5GpK6nEdU

10. Camille von Kaenel, "Utilities Are Giving People Cash for Clean Cars," *Scientific American*, July 17, 2017, https://www.scientificamerican.com/article/utilities-are-giving-people-cash-for-clean-cars/

11. Kristy Hartman and Emily Dowd, "State Efforts to Promote Hybrid and Electric Vehicles," National Conference of State Legislatures, September 26, 2017, http://www.ncsl.org/research/energy/state-electric-vehicle-incentives-state-chart.aspx

12. Mike Salisbury and Will Toor, "How Leading Utilities Are Embracing Electric Vehicles," Southwest Energy Efficiency Project, February 2016,

http://www.swenergy.org/data/sites/1/media/documents/publications/documents/How_Leading_Utilities_Are_Embracing_EVs_Feb-2016.pdf

13. Dana Sanchez, "Not the First in Africa: Cape Town to Use Electric Buses," AFK Insider, January 25, 2016, https://afkinsider.com/115166/not-first-africa-cape-town-use-electric-buses/

14. Douglas A. Bolduc, "China's BYD Plans EV Plant in Morocco," Automotive News Europe, December 9, 2017, http://europe.autonews.com/article/20171209/ANE/171209787/chinas-byd-plans-ev-plant-in-morocco

15. Nancy Cook, "Looking to Fund a Clean Energy Project? You Need a Green Bank," CityLab, July 17, 2014, https://www.citylab.com/equity/2014/07/looking-to-fund-a-clean-energy-project-you-need-a-green-bank/374611/

16. Jeremy Deaton and Mariya Pylayev, "Can Green Banks Turn Renewable Energy into a Financial Attraction?" ThinkProgress, May 17, 2016, https://thinkprogress.org/can-green-banks-turn-renewable-energy-into-a-financial-attraction-d5d721b284ec/

17. Connecticut Green Bank, http://www.ctgreenbank.com/

18. Matt Frades, "Clean Energy Banks Could Foster Private Investment in Charging Stations," C2ES blog, December 5, 2014, https://www.c2es.org/2014/12/clean-energy-banks-could-foster-private-investment-in-charging-stations/

19. "Electrify America Selects Greenlots to Develop Operating Platform to Manage $2 Billion Investment in Coast-to-Coast Network of High Speed Electric Vehicle Charging Stations," PR Newswire, January 23, 2018, https://www.prnewswire.com/news-releases/electrify-america-selects-greenlots-to-develop-operating-platform-to-manage-2-billion-investment-in-coast-to-coast-network-of-high-speed-electric-vehicle-charging-stations-300586373.html

20. Joshua S. Hill, "European Investment Bank Commits €4.3 Billion to Renewable Energy," CleanTechnica, July 19, 2017, https://cleantechnica.com/2017/07/19/european-investment-bank-commits-e4-3-billion-renewable-energy/

21. John Manning, "Banking and Clean Energy: A Blossoming Friendship," International Banker, October 11, 2017, https://internationalbanker.com/banking/banking-clean-energy-blossoming-friendship/

22. "Charging Infrastructure Is a $2.7 Trillion Barrier to Electric Cars," MIT Technology Review, October 11, 2017, https://www.technologyreview.com/the-download/609101/charging-infrastructure-is-a-27-trillion-barrier-to-electric-cars/

23. "Supercharger," Tesla Motors, n.d., https://www.tesla.com/supercharger

24. Bengt Halvorson, "Tesla Installing More Urban Chargers as BMW and Nissan Build Road-Trip Waypoints," *Car and Driver*, April 25, 2017, https://www.caranddriver.com/news/a15341580/tesla-installing-more-urban-chargers-as-bmw-and-nissan-build-road-trip-waypoints/

25. Amanda Silvestri, "Nissan to Build I-95 Fast-Charging Network Connecting Boston, New York and DC," *New York Daily News*, April 18, 2017, http://www.nydailynews.com/autos/news/nissan-evgo-fast-charging-network-new-york-boston-washington-dc-article-1.3069648

26. Amanda Silvestri, "BMW, Nissan Partner up to Expand DC Fast Charging Stations for EVs," January 25, 2017, *New York Daily News*, http://www.nydailynews.com/autos/news/bmw-nissan-partner-up-dc-fast-charging-evgo-expansion-article-1.2955123

27. Fred Lambert, "BMW and Nissan Partner to Build 174 More DC Fast-Charging Stations for Their Electric Vehicles," Electrek, January 24, 2017, https://electrek.co/2017/01/24/bmw-nissan-dc-fast-charging-stations-electric-vehicles/

28. Amanda Silvestri, "Nissan to Build I-95 Fast-Charging Network Connecting Boston, New York and DC."

29. "Partnerships Result in Up to 100 Electric Vehicle Charging Stations Across Parks," U.S. Department of the Interior National Parks Service, n.d., https://www.nps.gov/articles/evstations.htm

30. "Our Network," Ionity, n.d., https://ionity.eu/

31. Justin Worland, "Electric Vehicles Are Here. Now We Need to Figure Out How to Charge Them," *Time*, October 12, 2017, http://time.com/4979227/electric-vehicles-are-here/

32. "Colorado Electric Vehicle Plan: In Support of the Executive Order, Supporting Colorado's Clean Energy Transition," State of Colorado, January 2018, https://www.colorado.gov/governor/sites/default/files/colorado_electric_vehicle_plan_-_january_2018.pdf

CHAPTER 12: CONFLICT CHEMISTRY: RARE EARTH MINING AND RANGE ANXIETY

1. Doron Myersdorf, "Organic Compounds and Super-Fast Charging," YouTube, December 14, 2016, https://www.youtube.com/watch?v=HJ5ovebsqZQ&t=36s

2. Ingrid Lunden, "Quick-Charging Battery Startup StoreDot Gets $60M on $500M Valuation Led by Daimler," TechCrunch Network, September 14, 2017, https://techcrunch.com/2017/09/14/

quick-charging-battery-startup-storedot-gets-6om-on-5oom-valuation-led-by-daimler/

3. "Electric Vehicle Charging Guide," ChargeHub, n.d., https://chargehub.com/en/electric-car-charging-guide.html#publiccharging

4. "StoreDot Enters Mass Production of its Super-fast Batteries," TechTime, December 2, 2018, https://techtime.news/2018/12/02/storedot-3/

5. Ingrid Lunden, "Quick-Charging Battery Startup StoreDot Gets $60M on $500M Valuation Led by Daimler."

6. Jurica Dujmovic, "This Technology May One Day Replace Your Lithium-Ion Batteries," MarketWatch, October 24, 2017, https://www.marketwatch.com/story/this-technology-may-one-day-replace-your-lithium-ion-batteries-2016-10-24

7. "Global Lithium Ion Energy Accumulator Market—Forecast to Cross $40 Billion by 2022—Research and Markets," Business Wire, November 9, 2017, https://www.businesswire.com/news/home/20171109006233/en/Global-Lithium-Ion-Energy-Accumulator-Market--

8. John Petersen, "Tesla's First Decade Of Battery Pack Progress - Much Ado About Nothing," Seeking Alpha, August 25, 2017, https://seekingalpha.com/article/4101993-teslas-first-decade-battery-pack-progress-much-ado-nothing

9. Steve Hanley, "10 Myths About Electric Cars. Are Any of Them True?" Gas2, October 16, 2017, https://gas2.org/2017/10/16/10-myths-about-electric-cars-any-true/

10. Prachi Patel, "Simple, Energy-Efficient Recycling Process for Lithium-Ion Batteries," *IEEE Spectrum*, January 29, 2018, https://spectrum.ieee.org/energywise/energy/environment/simple-energyefficient-recycling-process-for-lithiumion-cathodes

11. Jim Motavalli, "Forget Lithium—It's Rare Earth Minerals That Are in Short Supply for EVs," CBS News MoneyWatch, June 19, 2010, https://www.cbsnews.com/news/forget-lithium-its-rare-earth-minerals-that-are-in-short-supply-for-evs/

12. Steve Hanley, "10 Myths About Electric Cars. Are Any of Them True?"

13. Bodo Albrecht, "How 'Green' Is Lithium?" Tech Metals Insider, December 16, 2014, http://www.kitco.com/ind/Albrecht/2014-12-16-How-Green-is-Lithium.html

14. Cécile Bontron, "Rare-Earth Mining in China Comes at a Heavy Cost for Local Villages," *The Guardian*, August 7, 2012,

https://www.theguardian.com/environment/2012/aug/07/
china-rare-earth-village-pollution

15. Amanda Kay, "6 Lithium-ion Battery Types," Lithium Investing News,
February 6, 2018, https://investingnews.com/daily/resource-investing/
energy-investing/lithium-investing/6-types-of-lithium-ion-batteries/

16. "Fuji Pigment Unveils Aluminium-Air Battery Rechargeable by
Refilling Salty or Normal Water," PR Newswire, January 8, 2015,
https://www.prnewswire.com/news-releases/
fuji-pigment-unveils-aluminium-air-battery-rechargeable-by-refilling-
salty-or-normal-water-300017712.html

17. Rob Verger, "The U.S. Navy is designing safer batteries, because no one
wants a fire at sea," *Popular Science*, April 27. 2017, https://www.popsci.
com/us-navy-is-designing-safer-batteries-because-no-one-wants-fire-at-
sea#page-5

18. James Dunn, "Spongy Zinc Battery May Beat Lithium-Ion on Safety,
Price, Recycling," *North Bay Business Journal*, July 24, 2017, http://www.
northbaybusinessjournal.com/northbay/marincounty/7216363-181/
zinc-battery-vs-lithium-ion

19. "US Navy, EnZinc Develop Zinc Alternative to Lithium
Batteries," *Batteries International*, May 11, 2017,
http://www.batteriesinternational.com/2017/05/11/
us-navy-enzinc-develop-zinc-alternative-to-lithium-batteries/

20. Steve Hanley, "Battery Researchers Keep Coming Up
with New Breakthroughs," CleanTechnica, December
21, 2017, https://cleantechnica.com/2017/12/21/
battery-researchers-keep-coming-new-breakthroughs/

21. Steve Hanley, "EP Tender Is a Range Extender Trailer for Electric
Cars," ecomento, October 21, 2015, https://ecomento.com/2015/10/21/
ep-tender-electric-car-range-extender-trailer/

22. Jasmina Schmidt, "EP Tender: The Trailer That Helps Electric Cars
Cover Long Distances," *Reset*, March 4, 2017, https://en.reset.org/blog/
ep-tender-trailer-helps-electric-cars-cover-long-distances-04032017

CHAPTER 13: ROADS: HEY, ARE YOU GOING TO PAY FOR THAT?

1. "Funding Federal-aid Highways," U.S. Department of Transportation
Federal Highway Administration, Office of Policy and Governmental
Affairs, January 2017, https://www.fhwa.dot.gov/policy/olsp/
fundingfederalaid/07.cfm

2. "Funding Federal-aid Highways," U.S. Department of Transportation Federal Highway Administration.

3. "Millennials Outnumber Baby Boomers and Are Far More Diverse," U.S. Census Bureau, June 25, 2015, https://www.census.gov/newsroom/press-releases/2015/cb15-113.html

4. Michael Sivak and Brandon Schoettle, "Recent Decreases in the Proportion of Persons with a Driver's License Across All Age Groups," University of Michigan Transportation Research Institute, January 2016, http://www.umich.edu/~umtriswt/PDF/UMTRI-2016-4.pdf

5. "Highway Statistics 2017: Distribution of Licensed Drivers," Department of Transportation Federal Highway Administration, Office of Highway Policy Information, January 31, 2019, https://www.fhwa.dot.gov/policyinformation/statistics/2017/dl20.cfm

6. "3.2 Trillion Miles Driven on U.S. Roads in 2016," U.S. Department of Transportation Federal Highway Administration, Office of Public Affairs, February 21, 2017, https://www.fhwa.dot.gov/pressroom/fhwa1704.cfm

7. "Corporate Average Fuel Economy (CAFE) Standards," U.S. Department of Transportation, August 27, 2014, https://www.transportation.gov/mission/sustainability/corporate-average-fuel-economy-cafe-standards

8. "Gas Guzzler Tax Program Overview," U.S. Environmental Protection Agency Office of Transportation and Air Quality, September 2012, https://nepis.epa.gov/Exe/ZyPDF.cgi/P100F3YZ.PDF?Dockey=P100F3YZ.PDF

9. "DOT, EPA Set Aggressive National Standards for Fuel Economy and First Ever Greenhouse Gas Emission Levels for Passenger Cars and Light Trucks," U.S. Environmental Protection Agency, April 1, 2010, https://yosemite.epa.gov/opa/admpress.nsf/docf6618525a9efb85257359003fb69d/562b44f-2588b871a852576f800544e01!OpenDocument

10. "Corporate Average Fuel Economy (CAFE) Standards."

11. Paul Ausick, "Why Are There 115,000 (or 150,000) Gas Stations in America?" 24/7 Wall St., May 22, 2014, http://247wallst.com/economy/2014/05/22/why-are-there-115000-or-150000-gas-stations-in-america/

12. Clifford Atiyah, "The Tax Man Cometh: These 10 States Charge Extra Fees for Electric Cars," *Car and Driver*, December 3, 2015, https://blog.caranddriver.com/the-tax-man-cometh-these-10-states-charge-extra-fees-for-electric-cars/

13. Joseph Kile, "Testimony: The Status of the Highway Trust Fund and Options for Paying for Highway Spending," Congressional Budget Office, June 18, 2015, https://www.cbo.gov/sites/default/files/114th-congress-2015-2016/reports/50297-transportationtestimony-senate.pdf

14. "What Is the Highway Trust Fund, and How Is It Financed?" Tax Policy Center, n.d., http://www.taxpolicycenter.org/briefing-book/what-highway-trust-fund-and-how-it-financed

15. "What Is the Highway Trust Fund, and How Is It Financed?"

CHAPTER 14: PLASTICS AND FRIENDS: THE MARRIAGE OF EVS AND O&G

1. Jennifer Latson, "The Burning River That Sparked a Revolution," *Time*, June 22, 2015, http://time.com/3921976/cuyahoga-fire/

2. Michael Rotman, "Cuyahoga River Fire," Cleveland Historical Society, September 22, 2010, https://clevelandhistorical.org/items/show/63

3. "How We Use Energy: Transportation," The National Academies of Science, Engineering, Medicine, n.d., http://needtoknow.nas.edu/energy/energy-use/transportation/

4. "Oil: Crude and Petroleum Products Explained—Use of Oil," U.S. Energy Information Administration, September 28, 2018, https://www.eia.gov/energyexplained/index.php?page=oil_use

5. Jack Ewing, "France Plans to End the Sale of Gas and Diesel Cars by 2040," *The New York Times*, July 6, 2017, https://www.nytimes.com/2017/07/06/business/energy-environment/france-cars-ban-gas-diesel.html?_r=0

6. Ashley Cowburn, "UK Plans to Halt Production of Petrol Cars by 2040," *The Independent*, July 14, 2017, http://www.independent.co.uk/news/uk/politics/uk-petrol-diesel-cars-stop-production-2040-climate-change-plan-a7842056.html

7. Alanna Petroff, "These Countries Want to Ditch Gas and Diesel Cars," CNN Money, July 26, 2017, http://money.cnn.com/2017/07/26/autos/countries-that-are-banning-gas-cars-for-electric/index.html

8. "50 Fun Facts About Steel," American Iron and Steel Institute, n.d., https://www.steel.org/~/media/Files/AISI/Fact%20Sheets/50_Fun_Facts_About_Steel.pdf

9. Trefis Team, "Trends in Steel Usage in the Automotive Industry," *Forbes*, May 20, 2015, https://www.forbes.com/sites/greatspeculations/2015/05/20/trends-in-steel-usage-in-the-automotive-industry/#5d5c05931476

10. "Evolution of Materials in the Car Industry," Oxford Advanced Surfaces, n.d., https://www.oxfordsurfaces.com/resource/evolution-of-materials-in-the-car-industry/

11. "Lightweight Materials for Cars and Trucks," U.S. Department of Energy Office of Efficiency & Renewable Energy, n.d., https://energy.gov/eere/vehicles/lightweight-materials-cars-and-trucks

12. Stacy C. Davis, Susan W. Diegel, and Robert G. Boundy, "Transportation Energy Data Book: Edition 33," U.S. Department of Energy Center for Transportation Analysis, Energy and Transportation Science Division, July 2014, https://info.ornl.gov/sites/publications/Files/Pub50854.pdf

13. Canadian Plastics Industry Association "Plastics in Motion: Plastics Help Make Today's Cars Possible," Globe Newswire, January 28, 2016, https://globenewswire.com/news-release/2016/01/28/805705/10159602/en/Plastics-in-Motion-Plastics-help-make-today-s-cars-possible.html

14. "Electric Vehicle Plastics Market Worth 1.49 Billion USD by 2021," PR Newswire, October 28, 2016, https://www.prnewswire.com/news-releases/electric-vehicle-plastics-market-worth-149-billion-usd-by-2021-599002261.html

15. Taylor Muckerman, "Can Big Oil Survive the Threat of Electric Vehicles?" The Motley Fool, December 17, 2017, https://www.fool.com/investing/2017/12/17/can-big-oil-survive-the-threat-of-electric-vehicle.aspx

16. "Ethene (Ethylene)," The Essential Chemical Industry, January 4, 2017, http://www.essentialchemicalindustry.org/chemicals/ethene.html

17. Michael Schirber, "The Chemistry of Life: The Plastic in Cars," Live Science, May 26, 2009, https://www.livescience.com/5449-chemistry-life-plastic-cars.html

18. "Electric Vehicles Driving Demand for Plastics," Polymer Solutions, June 22, 2012, https://www.polymersolutions.com/blog/electric-vehicles-driving-demand-for-plastics/

19. "Frequently Asked Questions: How Much Oil Is Used to Make Plastic?" U.S. Energy Administration, May 17, 2017, https://www.eia.gov/tools/faqs/faq.php?id=34&t=6

20. Robert Brelsford, "Chevron Phillips Chemical Commissions Texas PE Units, Delays Cracker Startup," Oil & Gas Journal, September 20/2017, http://www.ogj.com/articles/2017/09/chevron-phillips-chemical-commissions-texas-pe-units-delays-cracker-startup.html

21. Taylor Muckerman, "Can Big Oil Survive the Threat of Electric Vehicles?"

22. ExxonMobil, "ExxonMobil Begins Production on New Polyethylene Line at Mont Belvieu Plastics Plant," BusinessWire, October 17, 2017, http://news.exxonmobil.com/press-release/exxonmobil-begins-production-new-polyethylene-line-mont-belvieu-plastics-plant

23. "How Much Oil Does It Take to Make One Car Tire?" BestUsedTires.com Blog, April 19, 2016, http://blog.bestusedtires.com/how-much-oil-make-one-car-tire/

24. Steven Ashley, "Tire-Makers Try Treading Lightly on the Environment," *Scientific American*, August 11, 2010, https://www.scientificamerican.com/article/greener-tires/

25. "Products Made from Petroleum," American Geosciences Institute, n.d., http://www.earthsciweek.org/classroom-activities/products-made-petroleum

26. "Carbon Black & Oilfield Crayons," American Oil & Gas Historical Society, n.d., https://aoghs.org/products/oilfield-paraffin/

27. "Lubricating Oil," How Products Are Made, n.d., http://www.madehow.com/Volume-1/Lubricating-Oil.html

28. Angie Schmitt, "Americans Are Driving Less, But Road Expansion Is Accelerating," StreetsBlog USA, February 20, 2015, https://usa.streetsblog.org/2015/02/20/americans-are-driving-less-but-road-expansion-is-accelerating/

29. "America on the Move," American Oil & Gas Historical Society, n.d. https://aoghs.org/transportation/america-on-the-move-smithsonian-exhibit/

30. "Asphalt Paves the Way," American Oil & Gas Historical Society, n.d., https://aoghs.org/products/asphalt-paves-the-way/

31. "Asphalt Paves the Way," American Oil & Gas Historical Society.

32. Andrew Ward, "Rise of electric cars challenges the world's thirst for oil," *Financial Times*, August 7, 2017, https://www.ft.com/content/3946f7f2-782a-11e7-a3e8-60495fe6ca71

33. Stephen Edelstein, "Which Oil Companies Invest in Renewables, Electric-Car Services?" Green Car Reports, February 6, 2017, https://www.greencarreports.com/news/1108717_which-oil-companies-invest-in-renewables-electric-car-services

34. Adam Vaughan, "Shell to Open Electric Vehicle Charging Points at UK Petrol Stations," *The Guardian*, October 17,

2017, https://www.theguardian.com/business/2017/oct/18/
shell-to-open-electric-vehicle-charging-points-at-uk-petrol-stations

35. Stephen Edelstein, "Which Oil Companies Invest in Renewables,
Electric-Car Services?"

PART IV: THE FUTURE
THE WORLD IN 2040

CHAPTER 15: FORECAST TROUBLES: THE NUMBER KERFUFFLE

1. William Barnett, "Four Steps to Forecast Total Market Demand,"
Harvard Business Review, July 1988, https://hbr.org/1988/07/
four-steps-to-forecast-total-market-demand

2. Prakash Loungani, "The Arcane Art of Predicting Recessions,"
Financial Times, December 18, 2000, https://www.imf.org/en/News/
Articles/2015/09/28/04/54/vc121800

3. Tim Harford, "An Astonishing Record—of Complete
Failure," *Financial Times*, May 30, 2014, https://www.ft.com/
content/70a2a978-adac-11e7-8076-0a4bdda92ca2

4. Elizabeth Leary, "Which Market Gurus Get It Right the Most?"
Kiplinger's Personal Finance, April 2013, https://www.kiplinger.com/
article/investing/T038-C000-S002-which-market-gurus-get-it-right-the-
most.html

5. "Carbon Tracker Initiative," Energy Transition Advisors, May 8,
2014, http://www.carbontracker.org/wp-content/uploads/2014/05/
Chapter1ETAdemandfinal.pdf

6. "Global EV Outlook 2017: Two Million and Counting," International
Energy Agency, 2017, https://www.iea.org/publications/
freepublications/publication/GlobalEVOutlook2017.pdf

7. Eric Brandt, "Denmark EV Sales Plummet with Tax Break Elimination,"
The Drive, June 2, 2017, http://www.thedrive.com/news/11089/
denmark-ev-sales-plummet-with-tax-break-elimination

8. David Yager, "Electric Cars Won't Bring Down Oil Prices Anytime
Soon," *Business Insider*, July 29, 2017, http://www.businessinsider.com/
electric-cars-oil-price-2017-7

9. Reda Cherif, Fuad Hasanov, and Aditya Pande, "Riding the Energy
Transition: Oil Beyond 2040," International Monetary Fund Institute
for Capacity Development , May 2017, https://www.imf.org/~/media/
Files/Publications/WP/2017/wp17120.ashx

10. William Barnett, "Four Steps to Forecast Total Market Demand."

11. "Study: 81% of Consumers Say They Will Make Personal Sacrifices to Address Social, Environmental Issues," Sustainable Brands, May 27, 2015, http://www.sustainablebrands.com/news_and_views/stakeholder_trends_insights/sustainable_brands/study_81_consumers_say_they_will_make_

12. Matthew J. Hornsey, Emily A. Harris, Paul G. Bain, and Kelly S. Fielding, "Meta-Analyses of the Determinants and Outcomes of Belief in Climate Change," *Nature Climate Change*, 6, February 22, 2016, https://eprints.qut.edu.au/93213/1/93213.pdf

13. Earl J. Ritchie, "Driving to Work Alone Is a Costly Habit, So Why Do We Keep Doing It?" *Forbes*, June 16, 2016, https://www.forbes.com/sites/uhenergy/2016/06/16/driving-to-work-alone-is-a-costly-habit-so-why-do-we-keep-doing-it/2/#70dbf01d6248

14. Jeff Nilsson, "Why Fallout Shelters Never Caught On: A History," *The Saturday Evening Post*, August 26, 2011, http://www.saturdayeveningpost.com/2011/08/26/history/post-perspective/protection-cold-war-americans-fallout-shelters.html

15. "The Future of Cars 2040: Miles Traveled Will Soar While Sales of New Vehicles Will Slow, New IHS Markit Study Says," IHS Markit, November 14, 2017, http://news.ihsmarkit.com/press-release/energy-power-media/future-cars-2040-miles-traveled-will-soar-while-sales-new-vehicles-

16. Earl J. Ritchie, "Driving to Work Alone Is a Costly Habit, So Why Do We Keep Doing It?"

17. Jess Shankleman, "Electric Car Sales Are Surging, IEA Reports," Bloomberg, June 7, 2017, https://www.bloomberg.com/news/articles/2017-06-07/electric-car-market-goes-zero-to-2-million-in-five-years

18. "Number of Electric Vehicles on Roads Reaches Two Million: IEA," Reuters, June 7, 2017, https://www.reuters.com/article/us-autos-electric/number-of-electric-vehicles-on-roads-reaches-two-million-iea-idUSKBN18Y0XM

19. "Global EV Outlook 2017: Two Million and Counting."

20. William Barnett, "Four Steps to Forecast Total Market Demand."

21. Reda Cherif, Fuad Hasanov, and Aditya Pande, "Riding the Energy Transition: Oil Beyond 2040."

22. Rebecca Matulka, "The History of the Electric Car," US Department of Energy, September 15, 2014, https://www.energy.gov/articles/history-electric-car; Daniel Strohl, "How Henry Ford and Thomas Edison

Killed the Electric Car," Jalopnik, June 16, 2010, https://jalopnik.
com/5564999/the-failed-electric-car-of-henry-ford-and-thomas-edison

23. "Model T," History, n.d., http://www.history.com/topics/model-t

24. Stephen Leahy, "Electric Cars May Rule the World's
    Roads by 2040," *National Geographic*, September 13,
    2017, https://news.nationalgeographic.com/2017/09/
    electric-cars-replace-gasoline-engines-2040/

CHAPTER 16: LET'S MAKE (UP) A DATE

1. John Siciliano, "Rise of Electric Vehicles Threatens
   Oil Industry," *Washington Examiner*, December
   12, 2017, http://www.washingtonexaminer.com/
   rise-of-electric-vehicles-threatens-oil-industry/article/2642853

2. Allison Jones, "Ontario Has 'Zero' Chance of Meeting
   2020 Electric Vehicle Target, Analysts Say," CBC News,
   December 3, 2017, http://www.cbc.ca/news/canada/toronto/
   ontario-electric-vehicle-target-2020-1.4430760

3. John Morland, "Comparing the Top 5 European Countries for Electric
   Vehicle Adoption," FleetCarma, February 16, 2017, https://www.
   fleetcarma.com/european-countries-electric-vehicle-adoption/

4. Thomas Nilsen, "Electric Cars Set Sales Record in Norway with
   58% Market Share," *The Barents Observer*, April 1, 2019, https://
   thebarentsobserver.com/en/life-and-public/2019/03/60-all-new-car-
   sales-norway-are-now-pure-electric

5. David Jolly, "Norway Is a Model for Encouraging Electric Car
   Sales," *The New York Times*, October 16, 2015, https://www.nytimes.
   com/2015/10/17/business/international/norway-is-global-model-for-
   encouraging-sales-of-electric-cars.html

6. Sukanya Mukherjee, "India Gets Its Very First Electric Vehicle Charging
   Station in Nagpur," INC42, November 20, 2017, https://inc42.com/
   buzz/electric-vehicle-charging-station-nagpur/

7. "Scheme for Faster Adoption and Manufacturing of (Hybrid &) Electric
   Vehicles in India—FAME India—Reg. (Gazette Notification)," Ministry
   of Housing and Urban Affairs, Government of India, n.d. https://
   smartnet.niua.org/content/dc1659a0-3209-436f-a286-934651b4d9af

8. Chanchal Pal Chauhan, "Hail the Hybrid," *Business Today*, June 5, 2016,
   https://www.businesstoday.in/magazine/features/hybrid-vehicles-seem-
   to-work-than-the-purely-electric/story/232539.html

9. Greg Ip, "Electric Cars Are the Future? Not So Fast," *The Wall*

Street Journal, July 12, 2017, https://www.wsj.com/articles/
electric-cars-are-the-future-not-so-fast-1499873064#_=_

10. Steve Hanley, "EV Subsidies Going Up In Italy, Down In
    China," CleanTechnica, January 3, 2019, https://cleantechnica.
    com/2019/01/03/ev-subsidies-going-up-in-italy-down-in-china/

11. Mark Kane, "In April 2019, Plug-In Electric Car Sales In Italy Shot
    Up 271%," CleanTechnica, May 15, 2019, https://insideevs.com/
    news/349102/april-2019-plugin-sales-italy/

12. John Morland, "Comparing the Top 5 European Countries for Electric
    Vehicle Adoption."

13. Rebecca Linke, "The Real Barriers to Electric Vehicle
    Adoption," Newsroom, MIT Management Sloan School,
    August 3, 2017, http://mitsloan.mit.edu/newsroom/articles/
    the-real-barriers-to-electric-vehicle-adoption/

14. John Morland, "Comparing the Top 5 European Countries for Electric
    Vehicle Adoption."

15. Estefania Marchán and Lisa Viscidi, "Green Transportation: The
    Outlook for Electric Vehicles in Latin America," BMW Working
    Energy Paper, October 2017, https://www.thedialogue.org/wp-content/
    uploads/2015/10/Green-Transportation-The-Outlook-for-Electric-
    Vehicles-in-Latin-America.pdf

16. Anand Murali, "This Is Why India's Plan to Put 6 mn Electric Vehicles
    on the Road Will Fail," FactorDaily, June 29, 2017, https://factordaily.
    com/india-electric-vehicle-charging-mission-2020/

17. Rebecca Linke, "The Real Barriers to Electric Vehicle Adoption."

18. "Electric Cars: Will SA Switch On?" The Big Issue, n.d. http://www.
    bigissue.org.za/feature/electric-cars-will-sa-switch-on/

19. Norimitsu Onishi, "Weak Power Grids in Africa Stunt Economies
    and Fire Up Tempers," *The New York Times,* July 2, 2015, https://www.
    nytimes.com/2015/07/03/world/africa/weak-power-grids-in-africa-
    stunt-economies-and-fire-up-tempers.html

20. "Cars," Goldman Sachs.com, n.d., http://www.goldmansachs.com/
    our-thinking/technology-driving-innovation/cars-2025/

21. "Is There a Demand for Electric Cars in Africa?" *Business Chief,*
    August 28, 2016, http://africa.businesschief.com/technology/2516/
    Is-there-a-demand-for-electric-cars-in-Africa

22. Robert Ferris, "Electric Trucks Could Sell Faster than Cars, but Tesla
    May Be Aiming at the Wrong End of the Market," CNBC, November 21,

2017, https://www.cnbc.com/2017/11/21/tesla-semi-may-be-aiming-at-the-wrong-end-of-the-freight-industry.html

23. Jason Hall, John Rosevear, and Daniel Miller, "3 Major Challenges Facing Electric Vehicles That Few Investors Are Thinking About," The Motley Fool, September 12, 2015, https://www.fool.com/investing/general/2015/09/12/3-major-challenges-facing-electric-vehicles-that-f.aspx

24. Fred Lambert, "Lack of Awareness Is Surprisingly Still the Biggest Problem for Electric Vehicle Adoption," Electrek, January 3, 2017, https://electrek.co/2017/01/03/electric-vehicle-adoption-awareness/

25. Fred Lambert, "Automakers Spending on Ads for Electric Vehicles Show They Are Not Serious About Selling EVs," Electrek, December 21, 2016, https://electrek.co/2016/12/21/automakers-spending-ads-electric-vehicles-not-serious/

26. Andrew J. Hawkins, "Electric Cars Are Now Required to Make Noise at Low Speeds so They Don't Sneak Up and Kill Us," The Verge, November 16, 2016, https://www.theverge.com/2016/11/16/13651106/electric-car-noise-nhtsa-rule-blind-pedestrian-safety

27. Flavia Tsang, Janice S. Pedersen, Steven Wooding, and Dimitris Potoglou, "Bringing the Electric Vehicle to the Mass Market: A Review of Barriers, Facilitators, and Policy Interventions," RAND Europe Working Paper Series, February 2012, https://www.rand.org/content/dam/rand/pubs/working_papers/2012/RAND_WR775.pdf

28. "Ethanol Fuel in Brazil," Wikipedia, n.d., https://en.wikipedia.org/wiki/Ethanol_fuel_in_Brazil

29. "Lack of Electrification Strategy Could Hurt Three-Fourths of Top Auto Suppliers in Next Decade," PR Newswire, July 31, 2017, https://www.prnewswire.com/news-releases/lack-of-electrification-strategy-could-hurt-three-fourths-of-top-auto-suppliers-in-next-decade-300496020.html

30. David McHugh, "How Realistic Are Plans to Ban New Gas and Diesel Cars?" Toronto Star, August 4, 2017, https://www.thestar.com/business/tech_news/2017/08/04/how-realistic-are-plans-to-ban-new-gas-and-diesel-cars.html

31. Nick Carey, "UAW Presses Ford to Protect Factory Jobs Amid EV Push," Automotive News, October 5, 2017, http://www.autonews.com/article/20171005/OEM01/171009851/uaw-presses-ford-to-protect-factory-jobs-amid-ev-push

32. Steve Fekete, "Crude Oil and Refining Markets: Finding a New

Balance," HIS Energy, June 9, 2016, http://www.coqa-inc.org/docs/default-source/edmonton-2016/8-fekete.pdf?sfvrsn=d71147bb_4

33. Anindya Upadhyay and Iain Wilson, "Electric Vehicles Could Displace 8 Million Barrels of Oil Per Day by 2040," Bloomberg New Energy Finance, November 29, 2017, https://about.bnef.com/blog/electric-vehicles-could-displace-8-million-barrels-of-oil-per-day-by-2040/

34. David Koranyi, "Will Electric Cars Destabilize the World?" *The National Interest*, May 29, 2016, http://nationalinterest.org/feature/will-electric-cars-destabilize-the-world-16387

35. Ksenia Galouchko and Stephen Bierman, "Top Russia Oil Boss Scorns Tesla, Electric Cars as Overrated," Bloomberg, June 2, 2017, https://www.bloomberg.com/news/articles/2017-06-02/top-russian-oil-boss-scorns-tesla-electric-cars-as-overrated

36. "Only 50 Electric Cars Have Been Sold in Russia During the January-August Period," RusAutoNews.com, October 5, 2017, http://rusautonews.com/2017/10/05/only-50-electric-cars-have-been-sold-in-russia-during-the-january-august-period/

37. Stephen Edelstein, "Saudi Arabia Will Have $2 Trillion to Figure out an Economy After Oil" Green Car Reports, April 5, 2016, https://www.greencarreports.com/news/1103239_saudi-arabia-will-have-2-trillion-to-figure-out-an-economy-after-oil

38. Camilla Hodgson, "Saudi Aramco Execs See Uber as a Bigger Threat to Oil Demand than Tesla," *Business Insider*, December 11, 2017, http://www.businessinsider.com/saudi-aramco-ride-sharing-apps-than-electric-cars-2017-12

39. Anjli Raval and Andrew Ward, "Saudi Aramco Executives See Ride-Sharing as Threat to Oil Demand," *Financial Times*, December 10, 2017, https://www.ft.com/content/f0642f4a-dc20-11e7-a039-c64b1c09b482

40. Camilla Hodgson, "Saudi Aramco Execs See Uber as a Bigger Threat to Oil Demand than Tesla."

CHAPTER 17: THE FUTURE IS COMING

1. Melissa Grant, "Ipswich was Built on Mining But City Is Now Anti-Coal," *The Sydney Morning Herald*, August 26, 2015, https://www.smh.com.au/business/companies/qld-city-built-on-mining-is-now-anticoal-20150826-gj82fl.html

2. Hayden Johnson, "Electric, Self-Driving Cars to Shape City's Future: Tully," *The Queensland Times*, February 7, 2018, https://www.qt.com.au/news/electric-cars-to-shape-ipswichs-future-tully/3328475/

3. "Path to Autonomy: Self Driving Car Levels 0 to 5 Explained," *Car and Driver*, October 2017, https://www.caranddriver.com/features/path-to-autonomy-self-driving-car-levels-0-to-5-explained-feature

4. Justin Hughes, "Car Autonomy Levels Explained." The Drive, November 3, 2017, http://www.thedrive.com/sheetmetal/15724/what-are-these-levels-of-autonomy-anyway

5. "Audi Beats Tesla (And GM) to Level 3 Autonomy," Seeking Alpha, July 13, 2017, https://seekingalpha.com/article/4087480-audi-beats-tesla-gm-level-3-autonomy

6. Aarian Marshall, "After Peak Hype, Self-Driving Cars Enter the Trough of Disillusionment," *Wired*, December 29, 2017, https://www.wired.com/story/self-driving-cars-challenges/

7. Aarian Marshall, "After Peak Hype, Self-Driving Cars Enter the Trough of Disillusionment."

8. Karen Hao, "At Least 47 Cities Around the World Are Piloting Self-Driving Cars," Quartz, December 4, 2017, https://qz.com/1146038/at-least-47-cities-around-the-world-are-piloting-self-driving-cars/

9. Rachel Metz, "Apparently People Say 'Thank You' to Self-Driving Pizza Delivery Vehicles," *MIT Technology Review*, January 10, 2018, https://www.technologyreview.com/s/609944/apparently-people-say-thank-you-to-self-driving-pizza-delivery-vehicles/

10. Kevin O'Malley, "How Can a City Welcome Driverless Cars?" Prospect, September 10, 2017, https://www.prospectmagazine.co.uk/politics/how-can-a-city-welcome-driverless-cars

11. Greg Gardner, "Why Most Self-Driving Cars Will Be Electric," *USA Today*, September 19, 2016, https://www.usatoday.com/story/money/cars/2016/09/19/why-most-self-driving-cars-electric/90614734/.

12. Andrew J. Hawkins, "Not All of Our Self-Driving Cars Will Be Electrically Powered—Here's Why," The Verge, December 12, 2017, https://www.theverge.com/2017/12/12/16748024/self-driving-electric-hybrid-ev-av-gm-ford

13. Roberto Baldwin, "It Takes a Smart City to Make Cars Truly Autonomous," *Engadget*, June 14, 2017, https://www.engadget.com/2017/06/14/it-takes-a-smart-city-to-make-cars-truly-autonomous/

14. "Electronic Revolution & the Evolution of Future Automotive," *ELE Times*, January 22, 2018, https://www.eletimes.com/electronic-revolution-evolution-future-automotive

15. Antuan Goodwin, "Qualcomm's Inductive Charging Road Could Pave the Way to New EVs," CNET RoadShow, May 18, 2017, https://www.

cnet.com/roadshow/news/qualcomms-inductive-charging-road-could-change-the-way-we-build-evs/

16. "Automotive Solutions," WiTricity.com, n.d., http://witricity.com/products/automotive/

17. Abigail Fagan, "Israel Tests Wireless Charging Roads for Electric Vehicles," *Scientific American*, May 11, 2017, https://www.scientificamerican.com/article/israel-tests-wireless-charging-roads-for-electric-vehicles/

18. Julian Spector, "Blockchain-enabled Electric Car Charging Comes to California," Greentech Media, August 2, 2017, https://www.greentechmedia.com/articles/read/blockchain-enabled-electric-car-charging-california#gs.CJaX9I0

19. Ben Schiller, "Need Car-Charging Infrastructure? How About Peer-to-Peer and on the Blockchain?" *Fast Company*, August 22, 2017, https://www.fastcompany.com/40455969/need-car-charging-infrastructure-how-about-peer-to-peer-and-on-the-blockchain

20. Valery Komissarov, "Are Flying Cars the Future of Transportation or an Inflated Expectation?" TechCrunch, November 8, 2017 https://techcrunch.com/2017/11/08/are-flying-cars-the-future-of-transportation-or-an-inflated-expectation/

21. "Electronic Revolution & the Evolution of Future Automotive," *ELE Times.*

CONCLUSION

1. "Weiss Energy Hall," The Houston Museum of Natural Science, n.d., http://www.hmns.org/exhibits/permanent-exhibitions/energy-hall/

2. Karen Weintraub, "King Penguins Are Endangered by Warmer Seas," *The New York Times*, February 26, 2018, https://www.nytimes.com/2018/02/26/science/king-penguins-antarctica-climate-change.html